T0295169

Water and Wastewater Pipeline Assessment Technologies

Water and Wastewater Pipeline Assessment Technologies

Classification Systems, Sensors, and Results Interpretation

Justin Starr

CRC Press is an imprint of the
Taylor & Francis Group, an **informa** business

First edition published 2021
by CRC Press
6000 Broken Sound Parkway NW, Suite 300, Boca Raton, FL 33487-2742

and by CRC Press
2 Park Square, Milton Park, Abingdon, Oxon, OX14 4RN

Library of Congress Cataloging-in-Publication Data

Names: Starr, Justin, author.
Title: Water and wastewater pipeline assessment technologies : classification systems, sensors, and results interpretation / Justin Starr.
Description: First edition. | Boca Raton : CRC Press, 2021. | Includes index.
Identifiers: LCCN 2020026393 (print) | LCCN 2020026394 (ebook) | ISBN 9780367188450 (hardback) | ISBN 9780429198731 (ebook)
Subjects: LCSH: Water-pipes--Inspection--United States. | Sewer-pipe--Inspection--United States. | Sewerage--Inspection--United States.
Classification: LCC TD491 .S73 2021 (print) | LCC TD491 (ebook) | DDC 628.1/50287--dc23
LC record available at https://lccn.loc.gov/2020026393
LC ebook record available at https://lccn.loc.gov/2020026394

ISBN: 978-0-367-18845-0 (hbk)
ISBN: 978-0-429-19873-1 (ebk)

Typeset in Times
by Deanta Global Publishing Services, Chennai, India

Contents

Preface....................vii
Author Bio....................ix
Introduction....................xi

Chapter 1 Infrastructure and the Need for Condition Assessment....................1

An Abbreviated History: Ancient Times through the Modern Era....................1
Current State of Wastewater Infrastructure: Materials and the Need for Inspections....................11
The Case for Inspection and Condition Assessment....................20

Chapter 2 Visual Inspection Technology and Data Coding Systems....................23

Mechanically Simple Technologies: Humans, Push Cameras, and Pole Cameras....................23
Active Mobility: Crawlers....................30
Inspection Conditions: Clean to Inspect or Inspect to Clean?....................46

Chapter 3 Classifying and Analyzing Visual Inspection Data....................51

Roots of Condition Assessment: WRC's MSCC....................51
NASSCO PACP....................56
Software Packages....................62
Practical Issues with PACP Inspections: Data Quality....................63

Chapter 4 Quantifying Condition above and below the Flow: An Overview of Sonar and Laser Technologies....................69

Profiling Sonar....................70
Laser/Lidar....................80
Pipe Wall Measurements....................93

Chapter 5 Methods for Detecting Hydrogen Sulfide Gas....................97

Deployment Considerations....................106
Development of Corrosion Models....................109

Chapter 6 Leak Detection – Static Sensors and Acoustic Inspections....................111

Meters and Instrument Readings....................114
Acoustic Detection Technologies....................119
Acoustic Inspection Technologies....................122
Other Applications of Acoustic Technology....................127

Chapter 7 Other Leak Detection Technologies: Focused Electrodes, Gas Tracers, and Infrared Imaging 133

Satellite Systems 133
Gas Tracers 139
Focused Electrode Leak Location (FELL) 144
The Importance of Strategic Planning and Cost–Benefit Analysis 149

Chapter 8 Condition Assessment in Water Lines 151

Materials and Failure Modes 151
Visual Inspection 158
Destructive Testing 159
Nondestructive Testing Methods 160
Statistical Analyses 166
The Future of Waterline Inspection 167

Chapter 9 Manholes, Laterals, and Ancillary Structures 169

Manholes 169
Condition Assessment Approaches 177
Laterals 188
Other Structures 193

Chapter 10 Software Packages and Asset Management 197

Asset Management Overview 197
CCTV Management Platforms 201
Asset Management Planning Tools 205
Geographic Information Systems 209
Enterprise Asset Management 216

Chapter 11 Future Trends in Condition Assessment 221

5G and Connected Platforms 221
Big Data and Artificial Intelligence 226
Inspection Hardware 231
Virtual Reality and Augmented Reality 233
Advanced Autonomy 236
Smart Infrastructure 238

Chapter 12 The Importance of Condition Assessment 241

Appendix A: Sample Cleaning and CCTV Inspection Specification 247

Index 257

Preface

Condition assessment is quickly becoming a hot topic in the engineering community. In many areas, changing demographics, aging assets, and deferred maintenance have left municipal stewards facing the prospect of massive water and wastewater projects in the coming years: the American Water Works Association estimates that spending could top $1 trillion over the next 25 years, in addition to normal operation and maintenance spending.

Traditional methods would call for systematically replacing assets that are nearing the end of their design lives. Yet, civil engineers have learned much about *in situ* behavior of buried pipes over the past decades. While some 100+ year-old cast iron water mains may remain in excellent condition, prestressed concrete circular pipe from the 1970s can fail spectacularly and unexpectedly – well before designers intended. Even within a class of materials, degradation can depend heavily on local conditions – some sections of a reinforced concrete interceptor may show minimal corrosion, while others may display significant wall loss, due to localized differences in hydrogen sulfide buildup.

Inspection and assessment programs can provide a smart solution to this problem. Taking stock of buried infrastructure can help utilities directors understand which pipes are at risk of imminent failure and which pipes can continue to provide decades of service. Nobody wants a sinkhole or loss event, and yet many pipelines have undergone decades of service without any type of condition assessment.

While there are accepted standards for pipeline construction, the inspection industry has become something of the wild west – new and innovative technologies are continually entering the market, claiming better defect resolution, inspection speed, and ease of deployment, without peer-reviewed data to back up those claims. Blindly adopting new technologies can lead to disappointment, especially if rich technical specifications are not able to generate usable data – or even if an inspection technology just isn't the right fit for the question at hand.

This text seeks to help project engineers wade through vendor sales material by providing an agnostic and unbiased overview of different technologies that are currently on the market, and an analysis of future directions. It is not a cookbook or a replacement for engineering: it does not prescribe a technology for detecting corrosion in a 54″ reinforced concrete interceptor. Instead, it provides the reader with an intelligent overview of the core features of inspection technology – explaining some of the fundamental science where appropriate and linking technical approaches to data and analysis. After reading the chapter on laser and lidar, the reader will have a clear idea of the strengths and weaknesses of each, and know what types of rehabilitation questions would be well suited for each type of sensor.

It also provides an overview of specifications – highlighting those that are critically important and explaining why others should play a minimal role in selection. Vendor spec sheets and sales pitches can be truly convincing, but not all data points are created equal. After reading the chapters on CCTV technology, you will learn (and see) how a well-lit lower resolution video can be far more useful than a poorly lit clip with more than 5x the resolution.

The text would also be appropriate for continuing education for civil engineers, internal training programs, or even an undergraduate elective course. While it does not contain the mathematical rigor of many standard design texts, it provides ample opportunity for a discussion-based program of study that will map closely to many of the tasks newly minted engineers or EITs will be assigned when working with collection systems or distribution systems. Supplementing this text with hands-on practice using inspection devices or software analysis tools could greatly enhance the experience.

This work would not have been possible without the support and guidance of several key individuals: I would like to thank my wife, Liliane, for her support during months of research, writing, and rewriting materials, as well as my parents for encouraging me to get involved in the wastewater industry many years ago. I would also like to thank my students, who graciously accepted a smaller amount of homework assignments while I was preparing this text, and my former colleagues at RedZone Robotics who understand the importance of condition assessment programs.

Finally, I would like to thank the reader – not just for selecting this text, but for expressing an interesting in infrastructure inspection. The future of our water and wastewater systems will depend on a group of world-class professionals making the right decisions about repair, rehabilitation, and replacement programs. Doing this the right way will require hard work to seek out unbiased information, without being tempted to just recycle a vendor-provided specification. Thank you for your efforts, and know that your work is truly appreciated. If you find any errors or have suggestions for future editions, I'd love to hear your feedback.

Justin Starr
jstarr@ccac.edu

Author Bio

Justin Starr is an avid roboticist and entrepreneur. He has worked on different facets of robotic systems since 2007, including several defense programs for QinetiQ North America's Technology Solutions Group. He was also Chief Technology Officer of RedZone Robotics, Inc., a developer of multisensor inspection devices. Dr Starr currently holds six U.S. patents for innovative infrastructure inspection devices and is an Assistant Professor of Engineering Technologies at the Community College of Allegheny County, where he teaches multiple courses on mechatronics, microcontrollers, telecommunications, and industrial robotics. He is also a Senior Technical Advisor at PreScouter, Inc., advising corporate clients on advances and developments in robotic systems, and is a co-founder of Janus Robotics, Inc., a startup applying AI to disaster relief and recovery operations. Dr Starr earned his PhD and MS in Materials Science and Engineering from the University of Florida and his BS in Engineering Science from the University of Virginia. When he isn't working, Justin enjoys collecting older Volvo cars, homebrewing, and aquaponic gardening.

Introduction

Welcome! If you are reading this, you are about to embark on a journey deep into the world of water and sewer lines. Maybe you design them, maybe you crawl into them, maybe you are trying to budget for their repair and upkeep. Regardless of your relationship with buried assets, a thorough knowledge of condition assessment technologies can help you do your job more effectively.

Inspectors and field crews can understand what techniques they should adopt to ensure information is gathered during CCTV inspections. Field service engineers can understand how certain types of failures impact data collection and determine what is or isn't acceptable. Engineering partners can provide more accurate lifespan estimates, repair forecasts, and rehabilitation plans. City planners and public utility directors can make limited budgets stretch even further.

Truly understanding these technologies requires one to be a jack of many trades – it requires some understanding of hardware and electronics, some understanding of engineering principles, hydraulics and fluid flow, and – increasingly – an appreciation of the power of data. No background knowledge is assumed, though some basic arithmetic will be used to demonstrate concepts from time to time.

Some of the information in this text comes from equipment vendors – there is no other reliable source of specifications, high-quality photography, and sample data. That said, this book is independent and vendor neutral. Equipment that is discussed should not be taken as an endorsement, and specific brands that are omitted have not been chosen deliberately. Enough information will be provided for you to evaluate a piece of unknown equipment or data standard in terms of reference specifications that should be provided by any reliable vendor or manufacturer.

This book aims to build a layered understanding of inspection technology: first, some background for the state of infrastructure in the United States will be discussed. If you are working in this field, you will have an appreciation of the level of disrepair some assets have fallen into. If you are new to the industry, prepare for a surprise that will make you think twice about spending time on manhole covers or catch basins. This discussion will outline the need for comprehensive inspection programs, and may cause you to rethink the way you currently do things in your municipality or firm.

After that, different types of inspection hardware will be discussed – beginning with simple systems, and continuing on to more complex devices. Analog video cameras and defect coding will be discussed, followed by a detailed technical discussion of profiling sonar, laser, and gas sensors. New technologies like acoustic inspection, dielectric and satellite measurements will also be discussed, as will technologies specific to manholes and laterals. Links to smart infrastructure and the incorporation of smart manhole covers, flow meters, and other connected hardware will also be detailed.

Next, the book will outline the importance of data through a discussion of global positioning system (GPS) data collection and integration into geographic information systems (GIS). Specific asset management software packages will also be discussed – smaller tools like WinCan and larger enterprise tools like Maximo will be covered as standalone platforms as well as parts of an integrated whole. Different database formats and data structures will be explained to the extent that an understanding is useful for analysis – this is not a SQL textbook, but understanding why a platform might use that format is valuable to know.

Finally, the book concludes with several reference specifications for inspection work. If you are a contractor or service provider, analyzing these specifications may help you gain a competitive edge. If you are a municipality looking to put work out to bid, these documents can help you ensure free and fair competition. If you are currently working with a specification, stop and ask yourself where it came from, and if any revisions might improve the quality of the inspection data that you receive.

If you read this entire text from cover to cover, you will be in a strong position to design and implement a comprehensive inspection program. If you read sections of it, you will have a better understanding of the current state of various technologies and plenty of practical examples to rely upon. If you have comments or feedback, please share them with the author, and if you think you've identified an error, we would welcome that information, as well.

Justin Starr, PhD
Assistant Professor of Robotics, Community College of Allegheny County
Former Chief Technology Officer, RedZone Robotics, Inc.

1 Infrastructure and the Need for Condition Assessment

AN ABBREVIATED HISTORY: ANCIENT TIMES THROUGH THE MODERN ERA

Sewer and water lines are often thought of as modern conveniences in the United States as many rural areas of Appalachia continued to use outhouses and wells until the 1930s. It was not until electrification made it possible to pump water long distances that indoor plumbing became a reality for many Americans. Yet, in the rest of the world, this was not the case – piping systems have been found in ruins dating back to pre-Roman times.

This should not come as a surprise! When humans began to transition from tribes of nomadic hunter-gatherers and form cities centered around agriculture, their earliest needs centered around water. A supply of fresh water was needed for cooking, cleaning, and irrigation. A means of removing waste was critical for disease prevention. It follows that some of the earliest attempts at engineering involved constructing structures to solve these problems. While this is not a history textbook (see Walksi, Journal AWWA 2006), some important milestones will be discussed, as they directly relate to the state of our infrastructure today.

Civilizations able to channel gravitational energy were able to bring water into their homes and expel waste with ease. Many different cultures offer competing claims to the first water and wastewater systems. Ruins in Crete reveal that the Ancient Greeks were installing elaborate plumbing systems as far back as 1800 BC (Figure 1.1). Modern-day tourists gawk at Roman aqueducts and lead piping dating back as far as 312 BC. Amazingly, the great sewer or *cloaca maxima*, running through the heart of Rome, not only has portions surviving to the present day, but many sections have been integrated into the city's modern sewer system.

This triumph of engineering spread throughout the Roman empire – elaborate sewer and water systems survive in Italy, France and Britain. While the design and construction techniques did not benefit from computers or modern principles of design, many aspects of the system are impressive: smaller gradients throughout the system prevent the buildup of water pressure and excess erosion. Conduits were constructed above and below ground, and stepped cascades were used to introduce aeration. Siphons were even constructed to enable water to pass through depressions and valleys where arch construction would have been cost-prohibitive.

FIGURE 1.1 Ancient sewers in Knossos, Crete. The Minoan civilization had some of the earliest examples of water and sewer systems. Public Domain: https://commons.wikimedia.org/wiki/File:Knossos_sewers_PA067399.JPG

The Romans were also ahead of their time in their use of lead for piping considerations. Flexible, durable, and inexpensive, lead was a natural choice for piping applications. Lead also resisted developing cracks, pinholes, and other leaks, and could be easily connected segment by segment through soldering or brazing. It is also waterproof and relatively corrosion-free. The two concepts are so intertwined that the modern word "plumbing" even stems from the Latin word for lead: *plumbum*. Many scholars and historians have argued over whether or not chronic lead poisoning may have played a role in the decline and fall of Roman civilization or in displays of madness by Roman emperors and citizens (Figure 1.2).

The truth may be difficult to fully establish, even with two millennia of hindsight. While some Roman skeletons have been found with elevated levels of lead, it is important to note that Romans encountered lead in far more aspects of daily life than the water supply: lead was a key component in plates and cosmetics; many wealthy Romans even used lead acetate as an alternative to sugar – one of the first artificial sweeteners.

An analysis of surviving water lines suggests that very little lead leaching occurred. These gravity-fed systems carried water from the same sources for centuries, allowing chemical processes to form a protective oxide coating on the interior of these pipes. This oxide coating largely prevented widespread leaching of lead into the water supply. As publicly funded structures, many aqueducts benefited from central planning and proactive maintenance. Roman aqueduct managers avoided connecting acidic sources of water into the system, preserving that protective oxide layer – a lesson that the public works staff in Flint, Michigan, would have been wise to take note of.

FIGURE 1.2 Surviving Roman lead pipe used to supply water to the public baths. This example is located in Bath, England, showing both the breadth of the Roman Empire and the longevity of their engineering. Photograph by Andrew Dunn. Creative Commons: https://commons.wikimedia.org/wiki/File:Lead_pipe_-_Bath_Roman_Baths.jpg

Unfortunately, the fall of Rome led to an abandonment of many engineering techniques throughout the Middle Ages. It wasn't until industrialization and urbanization led to outbreaks of typhoid and cholera that countries invested in updated piping systems. Many European cities took in drinking water from the same rivers to which waste was discharged – making it difficult for residents to escape waterborne disease. Prior to the construction of sewers, residents who did not simply dump waste into gutters along the street made use of large cesspits – chambers under floors or in yards into which waste was regularly dumped.

Cesspits were not watertight structures – nor were they designed to be. Over time, liquid would seep out of the sides and bottom of the structure, leaving behind solid waste. This liquid seepage often contaminated groundwater, and the built-up filth remaining in the cesspits had to be emptied every few years. Nightmen or "gong farmers" had the unenviable job of entering and cleaning out cesspits, loading excrement into carts for disposal at predetermined locations, ranging from dumps to fields for fertilizer use.

As the prevalence of disease grew with urban populations, in the 1700s and 1800s, major engineering activities commenced to bring water into private homes and connect individual buildings to shared sewer systems for the disposal of waste. Sometimes these undertakings were driven by emperors: Napoleon commissioned the construction of 30 km of sewers in Paris. Other times, influential engineers spread the evangelism of water and sewer systems: British engineer William Lindley and his sons designed and constructed systems in Hamburg, Warsaw, St. Petersburg, Prague, and Budapest.

Perhaps the most well-known designer of sewerage systems was the British civil engineer, Joseph Bazalgette, who was responsible for the design and construction of a major network of interceptors (large diameter sewers that "intercept" flow from smaller upstream pipes) in London, designed to channel waste to the Thames estuary, far from drinking water supplies. Bazalgette's design incorporated more than 100 miles of interceptors fed by an additional 400+ miles of feeder systems. Waste that needed to be discharged upstream of the estuary was fed into treatment plants, and pumping stations allowed sewage to flow through regions ill-suited to gravitational design (Figure 1.3).

Bazalgette commenced his efforts with broad public support from both government and citizens. Recent legislation outlawing cesspits in London required buildings to connect to sewer systems – nearly all of which drained to the Thames. The result was a massive increase in sewage discharges, much of which was in close proximity to the intakes for water supply. This led to a series of cholera epidemics

FIGURE 1.3 The Victoria embankment was constructed to allow interceptors to collect sewage and prevent discharge to the Thames. This graphic showing construction is from the Illustrated London News, 1867. Public Domain: https://www.museumoflondon.org.uk/application/files/7315/5360/9930/Cross-section-of-the-Embankment.jpg

which killed tens of thousands of Londoners in 1848 and 1853. In 1858, an unseasonably warm summer, coupled with a prolonged drought, concentrated sewage in the Thames. This produced an additional outbreak of cholera, coupled with the "Great Stink" – a smell of sewage so foul that it spread throughout the city. In the days before germ theory was widely accepted, this stink was not just an inconvenience, but rather a source of panic, as residents feared that inhaling the smell could cause them to contract cholera or other diseases (Figure 1.4).

Thus, British society was receptive to Bazalgette's proposal for comprehensive water and sewerage systems. In addition to being functional, these systems were a testament to Victorian design sensibilities. Pumping stations from the era have decorative, elaborate ironworks, and tourists today can descend into the system, wade through waist-deep sewage, and marvel at how this Victorian-era system continues to serve London to the present day. Bazalgette's design not only effectively eliminated cholera in the city, but his idea to double capacity estimates allowed his structures to endure into the 21st century, even as London continues to grow.

In the United States, the first sewer systems took shape around the same time. While early colonial settlements sometimes used sewers or drainpipes made of

FIGURE 1.4 Drawing of a "tosher" – an individual who entered sewers and searched for valuables among the excrement in Victorian-era London. Public Domain: From London Labour and the London Poor (1850, Mayhew) https://www.smithsonianmag.com/history/quite-likely-the-worst-job-ever-319843/

FIGURE 1.5 East Coast cities such as Philadelphia were some of the earliest developers of sewer systems in the United States. This photo shows an interceptor under construction in 1883. Public Domain: http://www.phillyh2o.org/canvas/canvas07.htm

hollowed-out logs to drain waste from homes into nearby streams, the first comprehensive interceptor systems were constructed in the mid-1800s in cities like Boston, Chicago, and Brooklyn.[1]

American sanitation engineers – like Bazalgette – were attempting to design systems that could be future-proofed to some degree. Installing a large interceptor sewer was a major capital project (Figure 1.5). Cities investing in a system would not be thrilled at the prospect of digging new tunnels and boring through rock again in just a few years – as many engineers learned the hard way. In the absence of actual water usage data, trial and error was an unfortunate part of early sewer design, with backups and overflows becoming a common occurrence.

Critically, many of these engineers also made a decision that would have significant future consequences – they decided to kill two birds with one stone and design wastewater conveyance systems that not only took sewage to treatment plants, but also conveyed stormwater away from streets and surfaces during precipitation events. These combined sewer systems have become a major Achilles heel in America's buried infrastructure.

The idea behind combined sewer systems is logical: it doesn't make sense to install two complete sets of pipes for stormwater and sewage. In a completely separate system, the stormwater side will sit empty when there is no precipitation – something that applies to most days in places like Phoenix, Pittsburgh, and St. Louis. In a combined system, stormwater flows into the same interceptors that are carrying

[1] For more information on the history of sewers in America, see the excellent sewer history website compiled by Jon C. Schladweiler, http://www.sewerhistory.org.

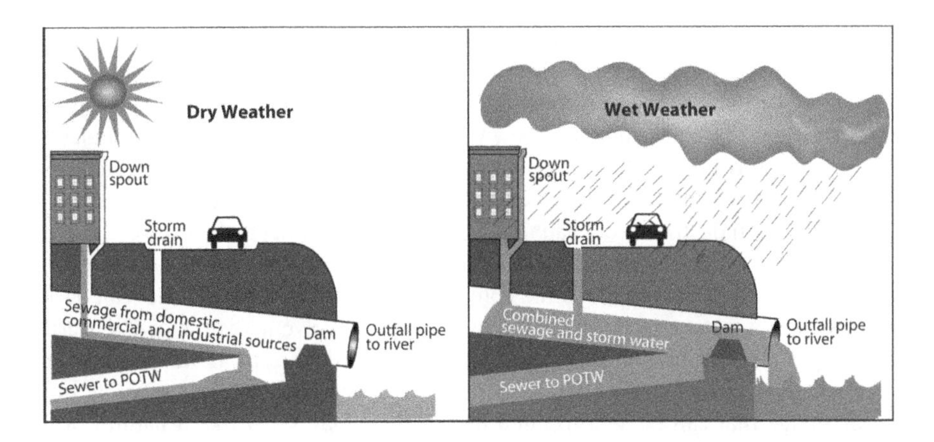

FIGURE 1.6 Combined sewers were commonly constructed before the passage of the Clean Water Act. This EPA image shows operation during dry weather and discharges during wet-weather events. Public Domain: https://en.wikipedia.org/wiki/Combined_sewer#/media/File:CSO_diagram_US_EPA.svg

sewage to the treatment plant. In the event that rainfall exceeds the capacity of the system, combined sewers were designed to intentionally overflow into surrounding waterways in order to prevent damage to the treatment plant (Figure 1.6).

While many of us recoil at the thought of intentionally discharging sewage into a waterway, at the time many of these systems were built, this was not so egregious. Until the 1970s, most wastewater treatment plants only engaged in primary treatment – debris were physically screened out of the system, solids were separated and allowed to settle, and effluent was discharged into waterways. This is a far cry from systems of today, and produced water that was visually clean, yet capable of harboring waterborne diseases and illnesses. Fecal coliform bacteria discharged in Pittsburgh can travel down the Ohio River, enter the Mississippi River, and impact users hundreds of miles downstream. To early engineers and designers, a few wet-weather discharges at strategic locations seemed more like smelly inconveniences rather than major threats to public health. At the time, industrial discharges meant that rivers and other waterways were not seen as desirable recreational destinations for the population. It was not uncommon for old cars to be placed along riverbanks as a method of erosion control – how picturesque!

America's waterways were so bad that they literally could catch on fire. In 1969, the Cuyahoga River was carrying significant amounts of debris from factories and industries in nearby Cleveland. In June of that year, a spark into the river caused the mixture of trash, chemicals, and other waste to ignite, causing a blaze that floated down the river. While the event did not cause loss of life, it indicated just how polluted rivers in the United States had become and was cited as one of the events that drove the passage of the Federal Water Pollution Control Act (or Clean Water Act) in 1972.

The Clean Water Act changed the regulatory landscape significantly, making it illegal for any person or entity to discharge a pollutant into a protected waterway.

While this targeted industrial users, it also applied to wastewater treatment plants – effluent from primary treatment was a major threat to public health and was classified as a pollutant, prohibited under the law. The Clean Water Act included significant funding for the construction of new wastewater treatment plants and upgrades to existing structures. For public infrastructure, the federal government would foot the bill for 75% of the capital cost, and states or local sources would be responsible for the remainder; industrial users had to pay for any improvements themselves.

Effectively, this ushered in an era of secondary treatment programs, which are designed to kill off bacteria, organic compounds and other materials that can harm organisms in a waterway. These expensive systems involve networks of UV sterilizers, digesters, clarifiers, and chlorination programs. The result is water that some treatment plant operators describe as being cleaner than potable, and a separate supply of sludge that can be sent to landfills. While it may seem like a natural fit to use this sludge for fertilizer on croplands, it is important to note that it may contain chemicals not suitable for human consumption – heavy metals, prescription drugs, and other harmful compounds.

While this program led to significant upgrades of wastewater treatment facilities around the country, it did little to improve the collection systems. Even if a plant doubled in capacity, interceptors that were already stressed to capacity during a wet-weather event were incapable of bringing all that flow to said plant. Redesigning the collection system was much more costly than plant upgrades – in many locations, effectively eliminating wet-weather overflows would have required the construction of entirely separate parallel systems, at costs of billions of dollars. Local legislative mandates meant that these costs would be borne by ratepayers, creating a situation where sewer bills could quickly increase to the size of mortgage payments. It should not come as a surprise that many municipalities put off these politically unpopular bond issues and upgrade programs. Continued combined sewer overflows meant that state and local governments were quickly becoming some of the biggest repeat violators of federal law.

The United States Environmental Protection Agency (EPA) turned to the court system for relief, suing major municipalities for knowingly failing to comply with the Clean Water Act. Expensive court cases are not in the best interests of either water authorities or the federal government, but such lawsuits indicated to municipalities the seriousness of the issue. In most cases, settlements were reached, with the EPA agreeing to forego prosecution in exchange for municipalities following detailed compliance programs – these settlements are signed off by the courts and are known as Consent Decrees (Figure 1.7).

Consent decrees have become significant issues in major metropolitan areas – wastewater treatment system operators in Boston, Pittsburgh, Scranton, Washington, DC, Baltimore, Honolulu, Atlanta, Chicago, Seattle, St. Louis, Toledo, and more are all operating under these court orders. An example of how these consent decrees impact local systems can be found by looking at any one of these cities.

In Pittsburgh, for example, the region's wastewater is treated by the Allegheny County Sanitary Authority (ALCOSAN). ALCOSAN's network of interceptor sewers, involving tunnels bored through solid rock at a depth of 100′ or more, was

THE UNITED STATES DISTRICT COURT
FOR THE WESTERN DISTRICT OF PENNSYLVANIA

UNITED STATES OF AMERICA,
COMMONWEALTH OF PENNSYLVANIA,
DEPARTMENT OF ENVIRONMENTAL
PROTECTION, and
ALLEGHENY COUNTY HEALTH
DEPARTMENT
Plaintiffs,

Civil Action No.

vs.

ALLEGHENY COUNTY SANITARY
AUTHORITY,
Defendant.

**UNITED STATES' NOTICE OF LODGING OF
PROPOSED CONSENT DECREE AND
MOTION FOR STAY OF LITIGATION**

The United States has lodged with the Clerk of Court a proposed Consent Decree which reflects a settlement agreement between the United States, the Commonwealth of Pennsylvania Department of Environmental Protection ("PADEP"), the Allegheny County Health Department ("ACHD"), and the Allegheny County Sanitary Authority ("ALCOSAN") in this action. If entered, this proposed Consent Decree would resolve the claims alleged in the complaint filed in this case.

FIGURE 1.7 Technically, combined and sanitary sewer overflows are violations of federal law. Consent decrees, such as the settlement shown above, are a tool that the EPA uses to get municipalities to comply with provisions of the Clean Water Act. Public Domain: https://www.epa.gov/sites/production/files/2013-09/documents/alcosan-cd.pdf

nominated for the "Outstanding Civil Engineering Achievement Award" by the American Society of Civil Engineers when construction was completed in 1959–1960. In the heyday of the steel industry, ALCOSAN's contribution to water quality paled in comparison to that of steel mills, chemical plants, and other heavy users of Pittsburgh's rivers. By the 1990s, most local industries had either left or cleaned up their operations, such that ALCOSAN was one of the larger contributors to river pollution in the region.

Facing a $275 million dollar fine, ALCOSAN entered into negotiations with the EPA, resulting in a consent decree that was finally signed off on in 2007–2008. As part of the terms of the decree, ALCOSAN paid a fine of $1.2 million, and committed to a plan to eliminate sanitary sewer overflows and significantly reduce combined sewer overflows by 85% by the end of 2026. The authority estimated that the costs of compliance would exceed $1 billion. Regional fragmentation meant that compliance started with mapping and locating systems that fed into ALCOSAN, routing rivers and streams out of the system, comprehensively inspecting lines to ensure capacity was being maximized, and installing a network of flow monitors in order to accurately respond to wet-weather events in real time. In the years around the consent decree, ALCOSAN ratepayers saw annual rate increases of 8.5–12%.

FIGURE 1.8 Cover page of ALCOSAN's Wet Weather Plan. At the time of this plan's release, the authority faced major criticism due to proposed rate increases and a lack of green infrastructure. Public Domain: https://www.alcosan.org/docs/default-source/clean-water-plan-documents/cwp-appendix-a/cwp-appendix-a-8a_wwp-companion-document.pdf?sfvrsn=481c9cf0_2

In 2012, ALCOSAN released a draft of a comprehensive Wet Weather Plan to eliminate sanitary sewer overflows and reduce combined sewer overflows by 96%. The plan involved construction of 25 miles of parallel interceptor infrastructure, significant treatment plant upgrades, and would cost rate payers more than $3.6 billion – funded by rate increases that would cause sewer bills to double every 7 years. The authority also proposed a less expensive $2 billion plan that did not fully meet EPA regulations but was seen as more sustainable (Figure 1.8).

Public outcry forced the authority to go back to the planning phases. In 2019, it received federal approval for a new plan that gave the authority an additional decade to achieve compliance, and incorporated green infrastructure – bioswales, rain gardens, permeable paving and distributed elements – designed to capture runoff and reduce inflow into the system during a wet-weather event. These alternatives are intended to reduce the need for extensive capital construction. Simultaneously, ALCOSAN is embarking on an expansive regionalization plan, where it will assume ownership and operational control of more than 200 miles of feeder sewer systems, enabling the authority to better control its ability to comply with federal requirements.

The story of ALCOSAN will continue in the coming decades. While the specific details of its saga and negotiations with the EPA, local stakeholders, and other interest groups are unique to the Pittsburgh area, other municipalities facing consent decrees offer variations on the theme: all are seeking to reduce or eliminate overflows. All are performing regular inspections in order to maximize available capacity, and all are struggling to deal with upstream influences not directly under their control.

At the same time municipalities have been working to comply with federal regulations, they have also been struggling to fund adequate operational and maintenance protocols, capital improvement programs, and other big ticket expenses go beyond federal compliance. How this lack of funding contributes to the state of sewers today is explored in the next section.

CURRENT STATE OF WASTEWATER INFRASTRUCTURE: MATERIALS AND THE NEED FOR INSPECTIONS

The previous section should give the reader a sense of appreciation for our current water and wastewater systems. In most developed nations, running water and sewage disposal are something that most are able to take for granted. While some areas still rely on septic systems and wells, that has become the exception rather than the rule. Yet, while the engineering that went into installing that infrastructure may be impressive, most municipalities have not maintained their buried assets with the same fervor that went into their installation. Note: While the degradation of sewer systems is a global issue, data presented in this section focuses on the United States.

To give a sense of the scale of the problem, in 2017, the American Society of Civil Engineers estimated that municipal governments spent approximately $30 billion on operation and maintenance of wastewater collection systems. Two years earlier, in 2015, Americans spent more than $33 billion on bottled water products, a figure that increased in subsequent years. The fact that Americans spend more for the convenience of bottled water than we do on maintaining our entire wastewater infrastructure should come as a shock: that $30 billion is spread across 800,000 miles of public sewers and wastewater treatment plants.

That number is grossly insufficient, but should not come as a shock within the context of America's infrastructure spending. In 2019, the American Road & Transportation Builders Association estimated that there were more than 47,000 bridges closed to traffic, in need of repair, or unable to carry the loads for which they were designed. They termed these bridges "structurally deficient" – indicating that while the bridges are not in danger of imminent collapse, they require monitoring, oversight, and vigilance in order to prevent public safety hazards, as these bridges continue to be crossed nearly 180 million times per year.

We can see these bridges decay before our very eyes. Just outside of Pittsburgh, Pennsylvania, the Greenfield Bridge became the archetype of this problem. Built in 1922, the Greenfield bridge was a concrete structure that carried traffic overtop of Interstate 376 – a major artery into the city. The National Association of Corrosion Engineers notes that most bridges from that era were constructed with a 50-year design life, meaning nobody should have been surprised when the Greenfield Bridge began to shed concrete in the 1980s. Motorists were injured by falling debris in 1989, and the city installed nets under the bridge to catch falling debris. When concrete eluded the nets and injured another motorist in 2003, the city installed a $700,000

FIGURE 1.9 Deferred maintenance and structures reaching the end of their design lives results in a significant amount of fixes and emergency maintenance. The former Greenfield Bridge in Pittsburgh had netting and a steel lower bridge installed to catch falling pieces of concrete. The bridge was demolished in 2015. Photograph by Doug Kerr. Creative Commons: https://commons.wikimedia.org/wiki/File:Interstate_376_-_Pennsylvania_(8461456052)_(2).jpg

bridge-under-the-bridge to prevent chunks of concrete from falling onto the roadway. The bridge was finally demolished and replaced with a new span in 2017 (Figure 1.9).

While many late-night television hosts poked fun at Pittsburgh's crumbling bridges, the consequences of failing to maintain these structures are no laughing matter. In 2000, the Hoan Bridge in Milwaukee, Wisconsin began to shift, buckle, and sag during morning rush hour. In Minneapolis, the collapse of the I-35W bridge killed 13 people in 2007, and in Italy, the sudden fall of a 210-meter section of roadway led to the deaths of 43 individuals. While sudden collapses are by definition unexpected, the deterioration of these structures is visible to the untrained eye, and crumbling infrastructure has become an accepted part of daily life in the United States. At current rates of funding, it would take nearly 80 years to repair all of the structurally deficient bridges across the country.

This suggests something even more ominous for water and wastewater systems. Many major water mains and sewage interceptors were constructed and installed at the same time as the bridges listed above. Yet, because they are buried out of sight, they receive even less attention from elected leaders, media groups, and everyday citizens. While the ASCE gave America's bridges an overall grade of C+ in 2017, it gave water and sewer systems grades of D and D+, respectively.

Water and sewer lines have undergone a shift in preferred material selection. Before 1900, most water lines were made from cast iron pipes – a material that was relatively inexpensive, strong enough for pressurized water lines, and capable of being buried under roads other structures. Cast iron is susceptible to corrosion but can be lined with cement to increase design life. While corrosion can cause external failures, the depositing of corrosion byproducts inside the pipe itself creates a phenomenon called tuberculation that can result in a significant loss of capacity. Cast iron pipes can also be used in sewer systems and face similar problems.

Cast iron pipes can be attacked externally as well – mostly through acidic soils (Figure 1.10). Damage to the pipe can result in significant water loss in a short period of time. When pipes are installed under roads, this can cause rapid washout of soil and the formation of a sinkhole. That said, cast iron remains a critical part of American infrastructure. Municipal Sewer and Water Magazine noted that in 2014, more than 600 municipalities had at least one cast iron pipe in service that was more than 100 years old. More than 20 systems had segments that were more than 150 years old.

The threat of sinkholes makes run to failure an unattractive option for cast iron pipe. A common replacement for cast iron lines is ductile iron pipe – made from iron mixed with 3–4% graphite. This extremely high carbon content enables ductile iron pipe to bend without breaking, giving it better fracture resistance than cast iron. To prevent corrosion from acidic soil, ductile iron pipe can be encased in polyethylene – an effective, but costly form of corrosion prevention. The ductility that gives ductile iron pipe its name also introduces a new failure mode: compressive forces can cause

FIGURE 1.10 Rust pits can be seen in the surface of this steel water pipe. This is not a new problem – the image above is from South Australia and was taken in 1925. Public Domain: https://commons.wikimedia.org/wiki/File:Rust_Pits_in_surface_of_Steel_Water_Pipe(GN04591A).jpg

crushing of the structure, and ovality – a phenomenon that can further weaken structural integrity and reduce system capacity.

Early large-diameter sewer lines were typically constructed from multiple courses of bricks. While construction of brick sewers was labor intensive, the refractory materials used to form bricks were naturally resistant to erosion and corrosion, though the same could not be said for mortar. Over time, loss of mortar could cause bricks to drop out, impeding the flow of sewage and presenting rehabilitative challenges – though fortunately, multiple courses provide some sense of redundancy. Further, compressive forces from the surface could cause ovaling of the brick sewer lines, and premature failure as modern development results in loads that would have been difficult to predict when these systems were designed and built at the beginning of the 20th century (Figure 1.11).

Smaller diameter lines were constructed from a variety of materials. In some places, logs were hollowed out and encased with steel straps in order to provide inexpensive gravity sewers. While these ideas seem crude to our modern sensibilities, wooden sewers remained in service in early industrial systems for decades longer than many thought due to a lack of reliable recordkeeping. In Spokane, Washington, a 2018 sewer rehabilitation project unearthed a section of wooden sewer still in use. Homes in the area were constructed from the 1880s to the 1910s, and the wooden line is estimated to date from that era.

While wooden sewers have been all but phased out of modern construction practices, vitrified clay is a material that has been used in sewers from ancient times to the modern era. Vitrified clay is formed by firing clay tubes in a furnace at temperatures exceeding

FIGURE 1.11 In sewers constructed of brick, loss of a few bricks can lead to accelerated degradation. In some sewers with multiple courses of brickwork, this may not be as significant as the image shown here. Soil and filler material is entering the pipeline unimpeded. Image taken from Sever et al., *Pipelines 2018*. STM (ASCE): https://ascelibrary.org/doi/pdf/10.1061/9780784481653.073

2000°C, causing the outer layers to transform into an amorphous glassy material that is watertight. Sometimes further glaze is applied to increase the impermeability to water. The process of making vitrified clay dates back thousands of years, and examples of these types of pipes have been found in Roman and Greek sewer systems.

Clay piping is also popular in the United States for many of the same reasons it was popular in ancient times. In areas with large clay deposits or soil containing clay, these pipes can be produced very inexpensively. Further, since they use locally sourced materials, they are considered to be extremely sustainable. The fragility and heavy weight of vitrified clay pipe makes transportation difficult, and as a result uptake and installation varies from region to region. Early vitrified clay pipes may be formed in a variety of nonstandard sizes, though modern lines commonly conform to standard diameters of 6–12″. Vitrified clay is not used in larger diameter lines due to the density and weight that would be inherent in pipes used in collectors or interceptors.

While vitrified clay was always known to be extremely tolerant of various materials flowing through the system, detailed modern tests have codified this. Vitrified clay is resistant to nearly all types of material found in residential and commercial sewage – acidic wastewater does not corrode the surface, and it is particularly resistant to hydrogen sulfide – a caustic chemical that will be discussed more in sections on concrete pipe. While seriously concentrated industrial wastes can damage vitrified clay, the biggest threat to the longevity of these sewer lines comes from physical forces. Improper installation and backfill can cause vitrified lines to break or shatter, just as illegal taps can lead to local collapses (Figure 1.12).

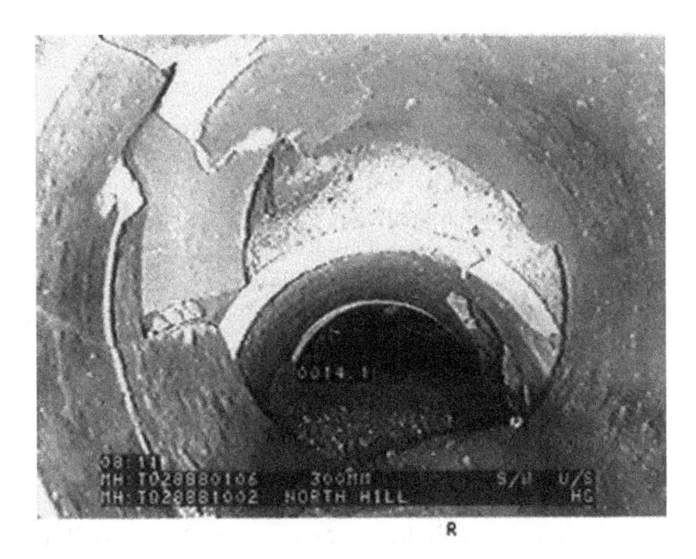

FIGURE 1.12 Photo of a collapsed rigid sewer line taken from a CCTV crawler. Non-deformable materials like vitrified clay and concrete often fail catastrophically. Failures like this can allow debris to wash into the sewer, and eventually create a sinkhole overhead. Image taken from Davies et al., *Urban Water.* STM (Elsevier): https://www.sciencedirect.com/science/article/pii/S1462075801000176#FIG1

Most vitrified clay pipes in the United States are structured with an interlocking bell and spigot design that prevents leakage so long as flow continues to progress downstream. Backups can lead to joint separation, and the fact that joints are not always perfectly sealed makes vitrified clay systems particularly susceptible to root intrusion. Roots that enter a vitrified clay system find a source of nutrient-rich water that can cause rapid growth. These root balls or masses can prevent the flow of wastewater and can lead to fracture or cracking of the surrounding pipes.

Modern materials science principles have been used to bring the benefits of polymers to water and wastewater lines. The same principles that make plastic an environmental nightmare also make forms of plastic well suited to applications in water and wastewater. The two main materials in use today are high-density polyethylene (HDPE) and polyvinyl chloride (PVC). Though crosslinked polyethylene (PEX) is commonly used in residential construction, it does not factor widely into municipal water and sewer lines and will not be discussed in this text.

HDPE has seen widespread use in the oil and gas and chemical processing industries, as it is inexpensive, lightweight, and fluid tight. HDPE can be joined together in segments using different types of plastic welding (butt welding, electrofusion, etc.) that create joints that are as strong as or stronger than the surrounding material. HDPE pipes have life estimates of 50–100 years, though much depends on the actual service conditions and installation methods used. Downsides of HDPE include the fact that the material is very temperature sensitive – losing strength and expanding as temperature increases. It also cannot be effectively glued, as no adhesives have been found that will maintain a seal as long as the surrounding pipeline. Further, it is subject to ovality, and many localities require extensive mandrel testing prior to acceptance. While cracks are not a common occurrence in HDPE, when they do occur, they tend to rapidly propagate, dramatically increasing the consequence of failure.

PVC is a much stronger material than HDPE, and thus requires much less material to achieve desired strengths. PVC pipes can be easily glued together using appropriate solvents and are stiff enough to be integrated into mechanical connections. PVC pipe is also resistant to common hydrocarbons typically found in wastewater – notably gasoline. That said, once installed, PVC pipes tend to offer years of fruitful service or fail rapidly due to improper installation. In the United States, PVC pipe is available in sizes ranging from 0.5 to 48″ and is one of the fastest-growing classes of materials used in new pipeline construction.

While other materials have been used over the years – including asbestos cement and pitch fiber (aka Orangeburg) – perhaps no material has had a bigger impact on both the growth of and issues facing modern sewers than concrete. Often thought of as a "miracle material," concrete has become a part of American life. Buildings, roadways, and piping systems all use concrete. Concrete can be easily poured into a mold and reinforced, or mass produced in precast structures. Concrete is the face of modernist architecture, as well as ancient buildings: both Le Corbusier and the Roman Colosseum used significant amounts of concrete.

When the growth of sewer systems exploded in the United States, concrete was the go-to material for large-diameter interceptors as it was thought to resolve many of the issues inherent to brick construction. First, concrete was much cheaper and faster to install than brickwork. Unskilled labor could be used to manipulate precast sections of concrete or pour mixtures into a mold. No expensive masons were required – a critical factor as many systems were installed during periods of high labor activity and unrest.

Further, concrete systems offered mechanical advantages over brick systems. When controlled for variations in thickness, concrete sewers could withstand higher compressive strength than brick sewers – a factor that was compounded by the ease of adding reinforcement to concrete pipes. In addition, it was easy to tailor concrete construction in response to on-site challenges with geography or topography – sewer diameters and thicknesses could be changed at any location by making slight variations in the mold. Custom connections to siphons or other structures could be easily fabricated as one-off creations – a fact that is noted in many as-built plans. The inner surface of concrete sewers could also be made smoother than brick sewers – a fact that engineers relied on to justify smaller diameter concrete pipes for given flow requirements.

Concrete structures also have extremely long design lives. Roman concrete used to construct the unreinforced dome of the Pantheon remains intact even today, and the U.S. Army Corps of Engineers recommends a design life of between 70 and 100 years for many different types of precast concrete piping sections – a fact that is cited by the American Concrete Pipe Association, which also notes many examples of concrete pipes being used for even longer service lives. Clearly, this statistic alone should cause collection system owners to take note: while the average age of water and sewer lines in the United States was only 45 years old in 2020, in some older cities like New York, this statistic more than doubles: in 2014, an average New York City sewer main was reported to be more than 84 years old! In 2015, DC Water reported that the median age of water and sewer lines in Washington, DC, was 79 years, with some pipes dating to pre-Civil War times. Even in newer cities like Jacksonville, Florida, many critical lines have already been in service for more than 65 years. The trend is clear: in many locations, large quantities of critical assets are rapidly entering the end of their useful lives.

Yet, the story is not so simple with concrete pipes. In reports dating back to the beginning of the 20th century, case studies noted that some concrete pipes experienced significant degradation just a few years after installation. In 1895, deterioration in a new outfall constructed in Los Angeles correlated to locations where flow traveled over a 12-foot drop. Even at the time, this enhanced decay was attributed to chemical attack, and a link between bacteria, sulfuric acid, and concrete sewers (as well as mortar between bricks) was becoming well known. In Metcalf's *American Sewerage Practice* from 1915, both the case of the Los Angeles Outfall and another case from Great Falls, Montana, are discussed.

The actual mechanism of this corrosion involves both chemical attack and biotic corrosion, and is summarized well by the Australian ARC Sewer Corrosion & Odour Research Linkage Process. In general, the microbial-induced corrosion process accounts for most corrosion present in concrete sewers. It typically takes place over three phases and results in predictable wall loss in concrete pipes (Figure 1.13). While evidence of this corrosive process was noted at the beginning of the 20th century, the full extent of the bacterial contribution has become a more recent phenomenon.

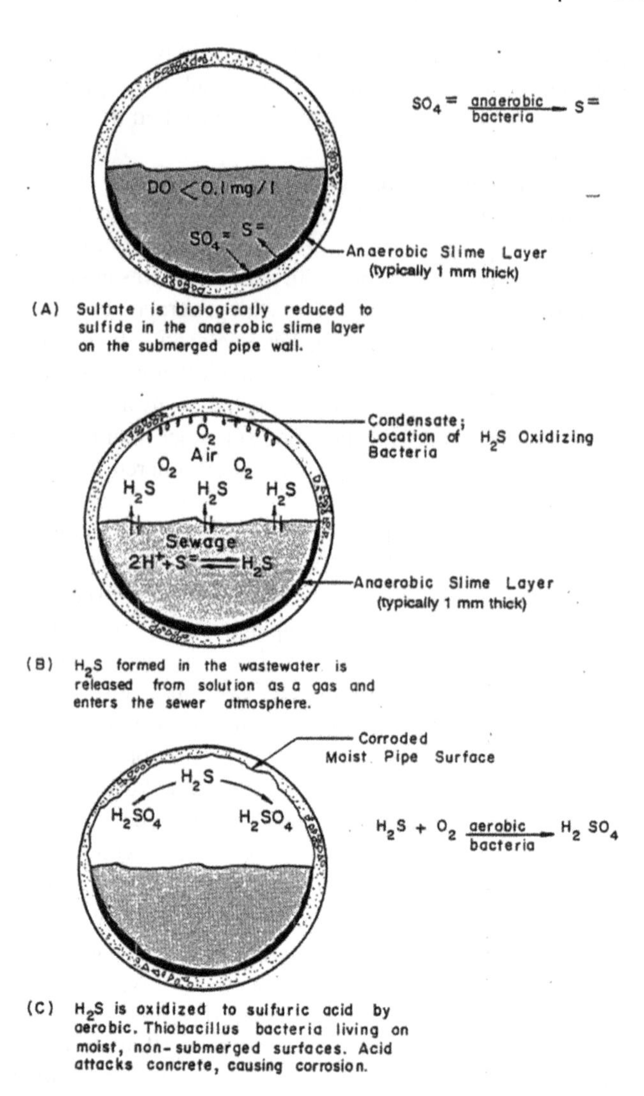

FIGURE 1.13 An illustration of the steps involved in the process of hydrogen sulfide corrosion of concrete pipes. This process has been observed since the early 1900s. Schematic provided by the EPA. Public Domain: https://nepis.epa.gov/Exe/ZyPDF.cgi/00000L69.PDF?Dockey=00000L69.PDF

The first phase of corrosion in concrete pipes is strictly chemical. The surface of a newly installed pipe (or indeed, any concrete structure) has a pH that is too high for any effective bacterial growth to take place. Naturally occurring bacteria (such as *Desulfovibrio desulfuricans*) in the wastewater stream break down sulfates and organic matter in order to release carbon dioxide and hydrogen sulfide gases. These gases diffuse into pores in the concrete structure. If moisture content is high, this can lead to the aqueous production of weak acids, which then react with underlying minerals like calcium hydroxide. It is important to know that this stage of corrosion does not actually lead to wall loss – instead, this process neutralizes the surface of the concrete, reducing the pH over the course of a few months.

The second stage of corrosion begins when the pH has dropped to the point where bacterial and fungal colonization is possible. In high humidity environments, these organisms (such as *Thiobacillus thioparus*) form a colony or slime on the surface of the concrete pipeline. They continue to break down hydrogen sulfide, carbon dioxide, and their byproducts, generating even more acidic materials, such as sulfuric and carboxylic acids. This stage of the corrosive process depends significantly on environmental conditions, but can take up to two years.

After this process completes, the pH of the wall of the pipe is now acidic, and bacteria begin to eat away at the concrete structure. Over time, concrete is transformed into other products, such as gypsum, which have little to no structural integrity. Once this mass loss begins, rate of corrosion will continue at a constant rate until the either the line fails, or corrective action is taken (assuming *in situ* conditions do not change). As corrosion takes place, scale and other byproducts occur at the surface, covering up activity. Unfortunately, this buildup does nothing to slow down the overall corrosive process, and in fact, can even mask the existence and rate of corrosion. Many asset owners report striking the side of healthy-looking pipe, and watching as large chunks or sheets of compromised material fall off.

This hydrogen sulfide-catalyzed corrosion can reduce the service life of a pipeline by a significant amount. The examples cited at the beginning of the 20th century saw crippling structural corruption that required remediation after just a few years in service. This form of corrosion only occurs in the headspace of flow; hence fully charged pipes are not susceptible. Further, corrosive activity increases with hydrogen sulfide content in the headspace, so well-ventilated designs can do much to increase asset life. Further, drops, eddies, or other flow disruptions can release dramatic amounts of gas, making localized corrosion likely and worrisome. Biological activity also increases with temperature and humidity – conditions which can be difficult to regulate.

Thus, the concrete pipe infrastructure in the United States is at a critical inflection point: many assets are approaching or exceeding their design lives, and harmful hydrogen sulfide corrosion is rapidly accelerating this process. Demographic changes have reduced urban wastewater flows so that assumptions about scouring velocity and vortices no longer hold true, while the capital budgets needed to sustain the growth of suburban systems take a toll on operations and maintenance.

The results can lead to catastrophe, and sinkholes caused by buried sewer and water lines are appearing at an increasing rate (Figure 1.14). In 2014, a sinkhole opened up under the rear tires of a fire truck in Racine, Wisconsin, trapping the unit in place. Engineers analyzing the incident directly attributed the results to hydrogen sulfide corrosion. In 2016, another sinkhole in Fraser Township, a suburb of Detroit, Michigan, condemned three homes and caused more than $75 million in damages. The cause? Premature corrosion of a concrete interceptor due to improperly performed repairs several years earlier. Pittsburgh again gained media infamy in 2019, when a sinkhole opened up under the rear wheels of a bus during morning rush hour, producing meme-worthy images that have given the city an image it would like to shake.

FIGURE 1.14 Sinkhole in a hotel parking lot in Georgia. While the sinkhole was centered under a parking lot, note the damage to buildings at the edge. The sinkhole was caused by the failure of a 120″+ combined sewer main. Image taken from Swanson and Larson, *Forensic Engineering*. STM (ASCE): https://ascelibrary.org/doi/pdf/10.1061/40482%28280%2913

While the Pittsburgh sinkhole is comical, other cases are no laughing matter. A 2016 sinkhole in San Antonio, Texas, opened up under a major roadway. Two vehicles and a pedestrian fell into the hole. One of the motorists died. The sinkhole opened up at the junction of an old concrete pipe scheduled for replacement and a new fiberglass line. So much soil was eroded due to the leakage that the sinkhole spread across the entire two-lane road. San Antonio suffered another major problem a few years later when in 2019, a sinkhole opened under busy U.S. Highway 90 – stemming from a failed 1985 cast iron pipe.

THE CASE FOR INSPECTION AND CONDITION ASSESSMENT

This epidemic of sinkholes coupled with the near-failing grades of D and D+ awarded to water and wastewater infrastructure should give those responsible for utilities pause: if Grade D infrastructure results in sinkholes, millions in damages, and loss of life, what will happen if things get worse? As public officials in Flint, Michigan, found, at some point courts may consider civil servants negligent to the point of criminal liability. Decades of conventional wisdom and assumptions are being stretched beyond the point of reasonableness – it is clear that something needs to change.

Unfortunately, massive increases in spending seem unlikely. In 2011, the EPA estimated that more than $384 billion in spending over two decades would be needed to maintain safe water quality. In 2012, the agency estimated an additional need of $271 billion to maintain clean watersheds through sewage treatment and associated programs. These estimates likely represent the tip of the iceberg, and more funding will be needed to proactively maintain these systems.

While the federal government provides some funding through the Clean Water State Revolving Fund ($1.4 billion annually), much spending on infrastructure takes place at the state and local level. Changing demographics have caused significant rate increases in some areas, such that improvements and repairs cannot be effectively funded by ratepayers. Financing through municipal bonds can leave cities saddled with debt, as the current generation of officials tries to cope with years of deferred maintenance. Even if we had the political will to do so, resolving this problem through spending would still take a significant amount of time.

It should come as no surprise then that inspection and condition assessment programs are rapidly gaining popularity. While municipalities have always inspected problematic lines, systemic inspection programs are becoming more and more pervasive. Some programs even use new technologies, including sonar, lidar, and other advanced sensors. The reason for this popularity is simple: as asset owners are forced to wrangle with massive underground infrastructure, investment programs help accomplish the following goals:

1) They help asset owners know exactly what they have – what lines are in the system and where are they located. Amazingly, a lack of recordkeeping means that many asset owners are unaware of the full extent or condition of their ownership. Witness the surprise in Spokane when a wooden sewer was encountered in 2018.
2) They help prioritize needed repairs. When many parts of a system are in need of overhaul, but there are insufficient funds to repair everything, a detailed analysis of condition can help establish what lines are at high risk of failure, moderate risk of failure, and low risk of failure – similar to an emergency medical triage system.
3) They can help make the case for more funding. Regular inspection programs generate large quantities of data. This data can be used to show the actual state of infrastructure condition and justify budget increases.
4) They can replace old, inefficient "rules of thumb." Some municipalities regularly replace sewer lines when performing major roadwork, reasoning that they don't want to dig twice. Inspection programs can create data-driven programs that replace assets when they really need replacing – not sooner than necessary.
5) They can provide a constantly changing view of a system. Buried infrastructure is interconnected in the same way that components of an organism are. Changes in one part of the system impact far-reaching extremities in strange and unusual ways. Comprehensive condition-assessment programs help asset owners understand and quantify these relationships.

Further, the EPA insists on detailed inspection programs as part of many consent decree settlements. Regular inspections of an entire collection system to ensure that capacity is being maximized are essential parts of a long-term control plan. Similarly, regular assessment of even smaller collection lines is critical to identify sources of infiltration or exfiltration, in addition to more serious problems.

As the demand for condition assessment programs increases, vendors are stepping in with new technologies that promise to save municipalities money and maximize useful life of infrastructure. Much of this discussion centers on detailed technical jargon and obfuscation rather than actual, tangible results – an insistence on quality is appropriate if it is necessary to generate useful inspection data – not to enable a single vendor to become "specced into" a particular contract.

2 Visual Inspection Technology and Data Coding Systems

Visual inspection of pipelines is one of the most common techniques in use today. Producing images that often do not require interpretation, it is useful to workers, engineers and elected leaders and remains a staple of the trade. Anyone presented with an image of plant roots or cracks knows immediately what they are looking at. That said, the qualitative nature of visual inspection makes it difficult to integrate into quantitative, data-driven inspection programs. This chapter will discuss techniques used to perform visual inspections of buried pipelines – including technology and best practices. It feeds directly into the next chapter which discusses coding and classification methodologies.

MECHANICALLY SIMPLE TECHNOLOGIES: HUMANS, PUSH CAMERAS, AND POLE CAMERAS

Many problems impacting water and sewer lines are visible to the naked eye: obstructions like roots can block flow, cracks can lead to a loss of structural integrity, and improper connections can enable large quantities of unwanted infiltration. Yet, as these structures are buried underground, obtaining a visual inspection can be a challenging process. It follows that some of the earliest forms of condition assessment involved ways of enabling humans to see the inside of pipes in order to observe and quantify problematic areas (Figure 2.1).

In large diameter interceptors, this process is relatively straightforward. It is possible for an individual to walk or otherwise navigate through a pipeline, recording problems as they are encountered. In the early days of pipelines, this could have meant taking detailed notes and returning with a mason to correct problems. Now, defects are cataloged with a brief description and a photograph. If stationing is visible inside a pipeline, the defect can be easily located at the surface. Absent this, a tether and encoder can help localize defects along the chainage from the entry point.

While historically popular due to necessity, manned inspection is not used as frequently in today's systems as its simplicity might suggest. While it does not require the purchase of expensive equipment, it does require significant safety precautions in order to protect the lives of inspectors. All pipelines must be checked for toxic hydrogen sulfide gas prior to manned entry, and flows must be reduced to safe levels, requiring expensive bypass pumping. Further, manned inspections are impractical on smaller interceptors, and impossible on small diameter lines.

FIGURE 2.1 Manned entry for inspection and maintenance has a long history. In the 1840s, (before the availability of jetters), "flushers" would enter a sewer, temporarily block the flow with boards allowing the water to rise, and suddenly release the flow allowing some degree of scouring to take place. Public Domain: https://yalebooksblog.co.uk/2014/09/21/dirty-old-london-30-days-filth-day-13/

Safety concerns cannot be overstated. Many sewer workers believe that they know a system well, and are under pressure to accomplish many operations and maintenance tasks in a short period of time. It can be tempting to skip hydrogen sulfide testing and the installation of custom blowers, especially for short, routine manned entry projects. Time saving actions like these can be deadly: exposure to concentrated pockets of hydrogen sulfide – 500–700 ppm – can cause an individual to collapse within 5 minutes. At 700 ppm, one can be knocked unconscious within two breaths, and at 1000 ppm, death can be instantaneous. Buried pipelines can be dangerous spaces, and proper confined space entry protocols should always be followed (Figure 2.2).

Deaths are less common in the United States due to the rarity of manned inspection. In the 21st century, most inspections are performed by teleoperated cameras or robotic systems – a topic that will be discussed in great detail in the following

FIGURE 2.2 Some tasks today still require manned entry – especially deployment of certain types of rehabilitation and repair systems and inspection of difficult-to-access structures. Remember, workers regularly die in sewers from a variety of hazards. Always use proper confined space entry procedures. Image taken from Reigard and Jones, *ASCE Pipelines 2017.* STM (ASCE): https://ascelibrary.org/doi/pdf/10.1061/9780784480892.031

sections. In countries that still rely on manned inspections, death is a much more common occurrence. In India, where sewer divers or scavengers bear responsibility for inspecting, cleaning, and repairing many sewer lines, the death toll remains high. In the first six months of 2019, more than 50 people died in 8 of the country's 36 states. A government report found that more people die in sewers every year than do fighting terrorism (Figure 2.3).

That said, when implemented properly and in limited forms, manned inspection techniques can be a valuable part of a condition assessment program. In cases of new pipe installation, a manned walkthrough can be a valuable part of acceptance testing (in addition to a mandrel). Further, some structures like outfalls can have irregular topography and little to nonexistent flow. It can be cost-effective to have an inspector enter the outfall at the point of discharge during times of low flow and quickly survey the structure.

In some cases, especially in smaller diameter lines where manned entry was impossible, smaller structures were constructed in order to facilitate visual inspection.

FIGURE 2.3 Sewer diving is one of the most dangerous jobs in existence. The top figure shows a diver preparing to enter a sewer in the United States, circa 1912. The bottom photo shows a sewer diver in Chennai, India, circa 2003. Note the lack of PPE. Some Indian government statistics indicate that a sewer worker dies every five days. The Manual Scavenging Act and other efforts are attempting to lower this death rate. Public Domain (upper): https://www.flickr.com/photos/seattlemunicipalarchives/19535685660 Creative Commons (lower): https://commons.wikimedia.org/wiki/File:Man_in_filled_sewer.jpg

These "lampholes" are rarely used today, but they were large enough to enable the lowering of a lamp and mirror down into a small diameter sewer (8″ or so) in order to enable a worker to check for blockages and look for issues. Larger manholes have proven to be much more useful for cleaning and repair, and are the norm today. Even though manned entry into 8″ sewer lines will likely always be a physical impossibility for humans, it is much easier to insert machinery into a pipeline for maintenance from a lower elevation making manholes invaluable.

While it is undesirable to frequently rely on manned entry for pipeline inspection, visual imagery is extremely useful in assessing condition. Technologies that make it possible to insert a camera into a pipeline have become increasingly popular and range in complexity from simple boroscopes to fully autonomous robots.

FIGURE 2.4 Example of a typical zoom camera, which can be easily and quickly deployed through manholes. In addition to the imager, an adjustable stop to keep the camera out of the flow and several lights are visible. Image taken from Selvakumar et al., *J. Pipeline System. Eng. Pract.* STM (ASCE): https://ascelibrary.org/doi/pdf/10.1061/%28ASCE%29PS.1949-1204.0000161

The simplest forms of camera conveyance technology are passive devices that do not have any internal means of propelling a camera down a pipeline. Common examples of these consist of push cameras, pole cameras, and zoom cameras (Figure 2.4). Push cameras are akin to larger boroscopes, in which a camera and an array of light-emitting diodes are mounted on the end of a cable that is attached to a large reel. As the cable is extended, the camera is pushed throughout a sewer line. In many cases, a video display is attached to the reel for real-time viewing of camera video. A simple serial encoder enables the distance the camera travels to be automatically overlaid on the video display (Figure 2.5).

Push cameras are appropriate for inspecting small-diameter sewers – lines ranging from 3 to 12″ in diameter. This range is determined by physics: the camera unit cannot physically fit in smaller pipelines, while the lighting is insufficient for large diameter lines. Further, the height of the camera inside the pipe cannot be adjusted, so low flow conditions are essential to accurately observe as much of the pipe as possible. Modern push cameras can be obtained with 400′ of cable, and rigid strain reliefs that make it possible to navigate 90° bends. Further, many can be operated entirely by battery power, making for less surface disruption and easier traffic control.

The most significant downside to push cameras is the fact that the camera head cannot be effectively controlled during deployment. As cable is extended to drive the camera head forward, the operator has limited degrees of freedom with which

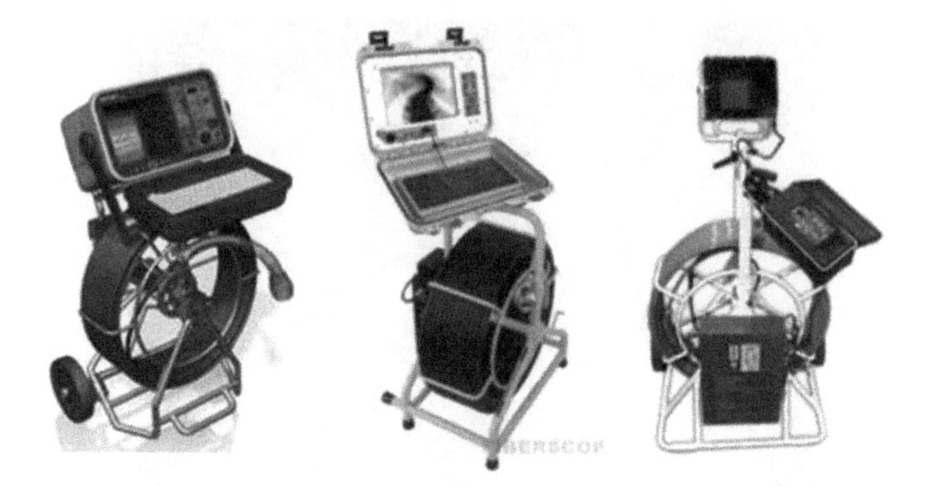

FIGURE 2.5 Several examples of push cameras. Advances in manufacturing and the ease of importation provide many options for a system. A unit from Envirosight or Cues (for example) comes with a warranty and support. Clone units can be imported directly from China for a fraction of the cost, but service and warranty support may be much more questionable. Image taken from Wade, *Pipelines*. STM (ASCE): https://ascelibrary.org/doi/pdf/10.1061/9780784413012.063

to pivot and orient the head itself and assert directional control. If the camera passes by a defect or lateral, it is difficult to pan or adjust the camera to get a better view. Further, careless users can easily drive a push camera up a lateral or wedge it in an offset joint. Much of this is offset by the low cost of these systems – a full setup can often be had for less than $5,000.

Most analog cameras output NTSC video format, while newer digital models generally output 640 × 480 pixel displays. While these may seem shockingly low in an era of 4K television, they are adequate for observing features in a 12″ pipeline. Push cameras come with adjustable lighting, and inspectors should make use of this feature. While it may slow down an inspection to dial in a well-lit image, cracks can be obscured in an image with too little or too much brightness. More expensive units feature built-in sondes for surface location and swiveling camera heads that can self-right during an inspection.

End users relying on push cameras as part of a comprehensive condition assessment program should use them appropriately. Push cameras are fine – and even desirable – as part of an emergency response kit, or part of a triage program. Need to see what is causing a blockage in a line? Insert a push camera in the nearest manhole and check things out. Want to quickly determine which lines would benefit from a more detailed subsequent inspection or take a statistical approach to cleaning? Quickly investigate the system with push cameras and make a list of targets for further follow-up. Systemic condition assessment using these technologies is not recommended.

Pole and zoom cameras operate according to similar principles. With these devices, a camera is attached to the end of a rigid pole, which can be inserted into a manhole or other access structure. Most poles are telescopic and can extend to distances of 20–30′ and can be swiveled to observe in multiple directions. The camera mounted on the pole often has a motorized tilting function, and can be zoomed in order to look down the length of a pipeline. This zooming functionality is so critical that many refer to pole cameras as zoom cameras.

Many pole cameras tout incredibly large zoom ratios – more than 400x! This is disingenuous. Many do offer 30–36x optical zoom capability coupled with a 10–12x digital zoom. These zooms are not created equal. Optical zooming relies on adjustments to the lenses in a camera system in order to effectively increase the magnification of an object. Both resolution and sharpness are maintained, so long as the image is focused. Digital zoom crops out a portion of the pixels on the screen, and resizes the image to fit. It does not enable the display of additional information, as cropping and resizing could easily be performed on a computer after the inspection. When purchasing a pole camera, special attention should be paid to the optical zoom functionality.

While older pole cameras generate NTSC video, newer devices have standardized on 1080p (HD) video streams – a resolution that is appropriate for inspecting all but the largest interceptors. More important than the resolution of the imager is the associated lighting. As light is required to produce an image, pole cameras rely on directionally focusing LEDs down the length of a pipeline. Many claim to be able to resolve objects 300′ away from the insertion point. While newer devices do have impressive lighting capabilities, usable imagery on most available pole cameras maxes out at about 100–200′ downpipe.

Further, while many vendors claim to have low-light imagers that can resolve features with as little as 0.001 lux of illumination available, more light will almost always yield a better image. Since many modern municipal codes specify that manholes on 8″ sewers should be no more than 500′ apart, it is usually fruitful to attempt to look at an obstruction or blockage from the other end of a pipe segment before resorting to a low-light image.

As pole cameras are not physically traversing a pipe segment, they are unable to make accurate chainage measurements. Many vendors have correlated zoom electronics with pipe sizes in order to estimate the distance of observed features, but this practice is less accurate than encoders and tethers. Sondes are also difficult to integrate, as the pole camera never moves horizontally into a pipe and does not truly enter the system.

A more significant downside to pole cameras is the fact that they provide even less ability to look at a defect or feature than the push camera. Observations of downpipe conditions are only obtained via zooming – it is impossible to pan or tilt to observe a crack in more detail. The intense LEDs required to obtain measurement at a distance can generate significant shadowing and other artifacts, and no amount of tilting or swiveling will enable a pole camera to see what is located behind a protruding tap 120′ from the manhole (or worse – to determine whether or not it is surrounded by void and the possibility of infiltration).

Just as with push cameras, pole and zoom cameras are effective tools for triage and emergency response. They enable rapid assessment of pipelines and require even less inspection time – and therefore less traffic control – than push cameras. They are also easy to transport and require much less clean up than push cams. In the case where a line may be blocked, a zoom camera inspection from the downstream end can quickly reveal if the cause is a root ball or collapsed pipe and enable corrective action to be quickly taken. The inability of pole cameras to interrogate observed features or defects means that they should not be used as comprehensive condition assessment tools.

When selecting a push, pole, or zoom camera, manufacturers' specifications may be of limited utility. Some are deliberately obfuscated in order to make it difficult to compare between comparable units. Other factors cannot be described accurately in quantitative terms. While megapixels certainly matter, the quality/brand of the internal imager and the precision of the lens can have just as much of an impact on the resulting video quality. For example, a Sony imager with a high-quality sapphire lens will produce much better quality than an offbrand clone imaging through polycarbonate – even if both are rated for 1080p.

The best way to make an informed decision is to try units from multiple vendors in your actual system. Many will allow an onsite demonstration of higher cost units. Choose a few of your problem lines with known defects and see which camera is able to provide the best representation of the issue. While trouble spots will vary from time to time, make sure your trials include at least a few of the same lines so that a comparison will be straightforward. Try to control for similar lighting conditions – if you cover a manhole for one brand, do it for the others as well. Let your crews try operating the control system – while iPad controllers are the fad of the day, sometimes large, physical plastic buttons are a better match for field work. Pretend to drop the controller. If the rep has a visible reaction, it probably isn't suitable for your crews.

High-quality units are available from major vendors, including Cues, Envirosight, Insight Vision, and Rausch. Consider all aspects of your operation when making a selection – do you intend to repair units on site or ship back for repair? Are parts shared with other pieces of equipment you may have already, like batteries and cables? Recently, several Chinese manufacturers have begun selling their own units – including Flexi and Wopson. These provide opportunities for significant savings – however, neither of these vendors have a significant North American service and repair infrastructure. While concerns about quality are understandable, they may not be isolated to these new names in the industry: many cameras available from domestic labels are manufactured on contract in China.

ACTIVE MOBILITY: CRAWLERS

Perhaps the most widespread tool in use to inspect underground lines is the teleoperated robot, or crawler system. These robots have tracks or wheels and are capable of driving down a pipeline, recording video en route. Many have an articulated head that is capable of panning, tilting, and zooming (PTZ) to better observe pipe conditions. This is the key feature that makes crawlers more useful than push or pole cameras: they are capable of not only entering a pipeline, but they can also be moved,

FIGURE 2.6 CCTV crawlers are teleoperated robots that are driven through a pipeline. This unit has four wheels, though others come with treads. Note the adjustable mast for centering the camera in the pipe. Public Domain: https://commons.wikimedia.org/wiki/File:Robot_zwiadowczy_z_kamer%C4%85_(Telewizyjna_Inspekcja_Sieci_Kanalizacyjnej_w_Toruniu).jpg

controlled, and manipulated with multiple degrees of freedom in order to capture the most useful views of items under observation. For example, a void in the pipe wall can be observed from multiple angles, to ensure it is not a shadow. Laterals can easily be examined, by panning the camera to look upstream into the connection. Rats and other vermin can be scared away with a few well-timed flashes of the lights on the front of the unit. Since all crawlers are designed to capture video data, they are commonly referred to as closed circuit television (CCTV) crawlers (Figure 2.6).

There is a wide variation in features associated with CCTV crawlers. Most are teleoperated and connected to the surface with a fiber optic or copper cable, but some have limited autonomy and are capable of driving from manhole to manhole without human intervention. Some feature special 360° cameras, or "fisheye" lenses, in order to capture more of the interior walls of the pipe without stopping to pan, tilt, and zoom – these are sometimes referred to as Sewer Scanning Evaluation Technology (SSET) cameras. Others have the ability to launch a portion of the robot up into laterals or connections to perform more detailed inspections. While the vast majority of CCTV inspection devices on the market today are indeed crawlers, there are some passive devices that use wastewater flow for propulsion. These floating systems are not crawlers per se, but still act to capture CCTV data. All crawlers have CCTV capability, but some also integrate other sensors such as sonar, laser, and hydrogen sulfide gas detectors.

This section will focus on the details of crawlers and CCTV capture. Since most inspection data reviewed by engineers, repair crews, and other decision makers consists of video or images, it is critical to understand what goes into providing high-quality inspection data. Some of this can be controlled through best practices, but some is tied closely to the features, specifications, and capabilities of the crawlers themselves. Later chapters in this text will explore some of the additional sensors that can be mounted on crawlers, and the associated costs and benefits of doing so.

Just as sewer and water lines come in all shapes and sizes, so too do crawlers. This is not a coincidence: in order to capture the best possible image quality, the video camera mounted on the crawler must be coincident with the centerline of the pipe. This is not just to appease symmetry enthusiasts – cameras that are too low or too high in a pipe are not the same distance away from features on all sides. For example, a camera that is above the centerline will be much closer to defects near the crown of a pipe, and farther away from defects in the invert. This can lead to preferential observation and missing key defects. Further, as the LED lights are often installed on, or in close proximity to, the camera head, this misplacement can lead to dramatically uneven lighting in the pipe – lighting so bad as to make the inspection unusable.

This can sometimes lead to counterintuitive results. A camera that is mounted too high will be closer to the crown of the pipe. In theory, this would mean that defects can be observed in more detail, as a larger quantity of pixels is available to capture the image. However, in practice, this is seldom the case, as the close proximity of LEDs to the crown of the pipe completely washes out the top portion of the video. Dimming the lights so that the crown is visible takes away from downpipe visibility, making a trade-off unacceptable. Asset owners commissioning inspection work should insist that cameras are centered during all CCTV inspections.

Another reason why poorly centered cameras can harm the quality of inspection data relates to physical traversal. Regardless of whether or not lines have been pre-cleaned, debris will be regularly encountered during the course of an inspection. If a camera is too high on the crawler, going over debris may cause it to gouge the crown of the pipe and can even end up in the robot becoming stuck. Similarly, if a camera is too low, it may not be possible to see over debris. Clever contractors may try to get away with using a CCTV crawler that is too small for the pipe in question by raising up the head a disproportionate amount. This can result in the crawler tipping over and becoming stuck, and usually fails to compensate for the small amount of lighting built into smaller crawlers.

Thus, it is imperative that CCTV crawlers be sized appropriately for the lines they are inspecting. This helps keep the camera centered, maintains video quality, prevents meandering from side to side in a pipe, and matches lighting with the available space. Sometimes access constraints and the conditions in unusual pipes will necessitate using a suboptimally sized crawler, but those decisions should be made with the full knowledge of the trade-offs in advance. Nothing should come as a surprise to the overall project manager.

This is critical and cannot be overstated. Whereas push cameras and zoom cameras are often employed for emergency work or brief surveys, CCTV crawlers are the workhorse of comprehensive condition assessment. Consulting engineers, utility managers, regulatory agencies, and the general public (albeit indirectly) rely on these devices to provide surveys of record. Not taking the time to choose the right crawler for an application could mean that critical information is missed, and will not be addressed for years to come. For example, many municipalities perform comprehensive inspections of all of their service lines at 5–10 year intervals – either due to internal policies or EPA consent decree requirements. If an improperly sized crawler with insufficient lighting fails to notice a void behind a lateral as part of a comprehensive inspection, that could mean 5 years of infiltration into the system, 5 years of washing out the soil above the connection, and 5 years for a dangerous sinkhole to develop, putting lives at risk.

In major condition assessment programs, it is critical to establish standards for video and inspection quality early on. If you are a municipality, consider sending an inspector to monitor crews during the initial stages of an inspection. Quickly review submitted data so that if there is a problem with an individual operator, you can address it quickly, before too many inspections have taken place. If you are a contractor, know how well your equipment matches the system, and think about the risks involved in cutting corners. While it might take your crews an additional 20 minutes to adjust the mast height on a crawler before commencing inspection, the cost of that adjustment is much cheaper than rework at the end of a project.

Because of this, crawlers come in a variety of shapes and sizes. Some crawlers have a modular base with a variety of wheel and camera mounting options that make one unit capable of inspecting a wide range of pipelines by quickly swapping modular options. Envirosight's RovverX system is capable of inspecting 6–24″ pipelines using this technique. Similarly, the Deep Trekker DT340 can accurately collect data from pipelines ranging from 8 to 36″. Some systems are not modular. RedZone Robotics produces the Solo inspection robot with a fixed body style. While the Solo has advanced autonomy, its configuration limits its utility to 8–12″ pipelines. Remember, just because a vendor's equipment is capable of inspecting a range of pipe sizes, actually doing so successfully will require additional configuration and should not be automatically assumed (Figure 2.7).

Lighting quality goes hand in hand with crawler sizing. Almost all crawlers available on the market today have adjustable LED lights built into, or in close proximity to, the camera unit. These LED lights can have their brightness controlled through physical buttons or automatic configuration. As noted previously, symmetric, even lighting is critical to a successful inspection. While equipment vendors will cite lumens and candle power, those raw numbers do not guarantee good lighting, and achieving a well-lit image is more of an art than a science.

Fortunately, one does not need to be a Picasso to achieve this. Start every inspection by varying the brightness of the LEDs. Adjust them so they are too bright, back off until they are too dim, and then land somewhere in the middle. Including this in the SOP or specification for an inspection will get crews to think about lighting every time they deploy. It can be tempting to simply use the settings from a previous inspection, especially if pipe diameter did not change. Coatings and interior conditions can have a dramatic impact on reflectivity and light absorption, making routine light adjustment a good practice to adopt (Figure 2.8).

It is also critical to adjust lights when encountering size or flow changes. If the flow level drops in a pipe, make sure the lighting is filling the available space now

FIGURE 2.7 It is possible for the functionality of a crawler to be replicated on a completely passive device such as a skid or floating platform. This example (from Poland) shows a CCTV skid that is pulled through the sewer via a winch and was taken by Leonard G in 2005. Public Domain: https://commons.wikimedia.org/wiki/File:PVIDownHole.JPG, https://commons.wikimedia.org/wiki/File:PVIVideoPig.JPG

that the reflective surface of the water has receded. If a crawler drives into a chamber, increase the lights while you look around the chamber. This represents a significant shortcoming of autonomous systems. While automatic lighting works well for photographers above the surface, underground computers still struggle to match the capabilities of humans. Detecting a loss in overall lighting and correcting this using an algorithm is computationally expensive, and would likely benefit from the use of graphics processor unit-specific code – a feature that is not available on autonomous systems currently in the field (as of this writing). Thus, the crude methods that are currently employed struggle with debris occlusion on lenses, grease films, water droplets, and more. With autonomous systems, these problems are often undetectable until the completion of an inspection – and are difficult to correct in the post-processing stage.

For municipalities, it is important to be aware of the fact that inspection providers will often attempt to justify using their fastest available technology to complete an inspection. They may ask you to approve using a certain robot in a certain inspection, even though it doesn't technically meet the project specification. Always ask to see a sample video from a previous deployment – in the same pipe size, flow conditions, and pipe material. Trust your eyes – if the video looks good and there are legitimate deployment challenges on the line in question, allow the substitution. However, if the contractor is asking to use a different robot on a 54″ concrete interceptor, and they can only show you video from a 36″ brick system, don't just assume lighting will scale. Make them use the right tool for the job. Consider allowing them to perform the inspection at risk, with no guarantee of acceptance – at best, they will succeed and avoid returning to a problematic line. At worst, they will generate a new sample video and better understand the capabilities and limitations of their own equipment (Figure 2.9).

Crawler size and lighting are two of the most important parameters to consider when it comes to generating high-quality video. Yet, inspection specifications and

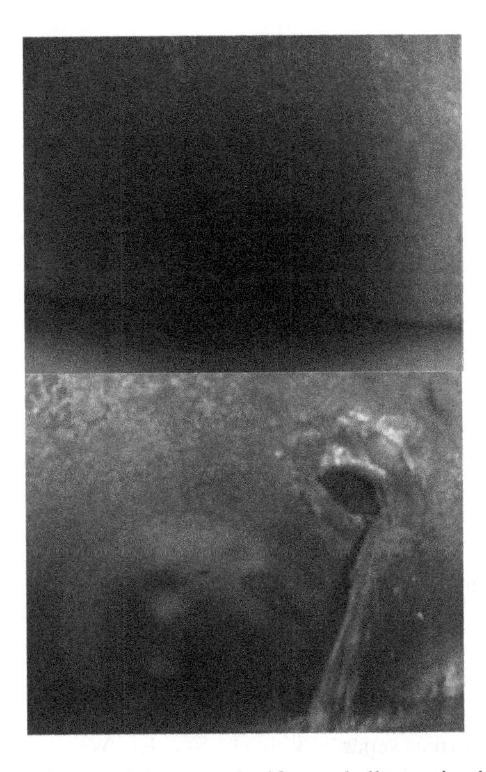

FIGURE 2.8 Lighting is one of the most significant challenges in obtaining high-quality CCTV data. The top image is taken from a floating platform in 66″ reinforced concrete pipe. Notice how difficult it is to see defects that are out of the well-lit periphery. The bottom image shows examples of water spots on the lens. Remember, these issues cannot easily be corrected in the post-processing stage. Public Domain: https://nepis.epa.gov/Exe/ZyPDF.cgi/P100C8E0.PDF?Dockey=P100C8E0.PDF

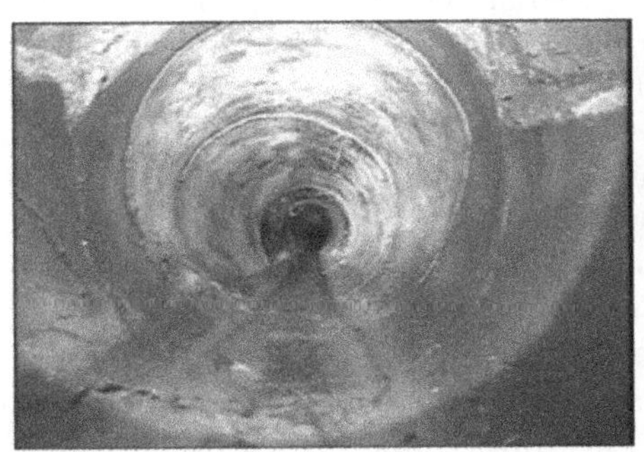

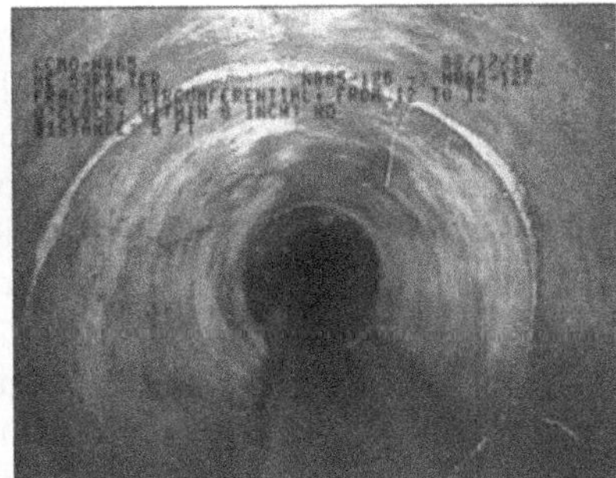

FIGURE 2.9 Side-by-side images of a zoom camera and traditional CCTV in the same pipeline. These images are taken from a 2011 EPA field demonstration. Note that in the demonstration, defect locations differed by 1–3′ between the technologies, and the zoom camera found far fewer defects than the CCTV inspection. Public Domain: https://nepis.epa.gov/Exe/ZyPDF.cgi/P100C8E0.PDF?Dockey=P100C8E0.PDF

vendor websites tout imager characteristics and file formats extensively. While these details are significant, basing inspection specifications on them or equating them with quality standards is a mistake. There are two main types of imagers, and each requires a slightly separate discussion – traditional PTZ cameras and 360° fisheye imaging systems.

Traditional PTZ cameras are relatively straightforward, and can be evaluated on many of the same metrics mentioned for pole and zoom cameras. There are two main types of PTZ cameras available today – analog and digital cameras. Analog cameras output standard definition video and follow the National Television System Committee or NTSC standard for color video. NTSC specifies 486 visible scan lines that are interlaced with one another, meaning the entire screen is drawn twice – once with even numbered lines, and again with odd numbered lines. This specification is sometimes referred to as 480i.

This technology was appropriate when images were being projected onto a phosphor screen, as decay of the brightness took time, and an image with effectively twice as much detail could be constructed in this way. However, digital technology means that this is less applicable and interlacing often leads to unpleasant-looking artifacts when panning quickly. This video can be stored on one, two, or three channels. When all of the video signals are mixed together into one channel, the output is known as composite, and can be recognized by the signature yellow RCA connector. When the video signal is split into two channels – one for brightness (or luminance, abbreviated by the letter Y), and the other for color (chrominance, C), the output is known as separate video or S-Video. When the output is split into three channels – one for Y, one for P_B (the difference between blue and luma), and one for P_R (the difference between red and luma) – the output is known as component video.

While all represent the same standard definition video signal, there can be a marked difference in video quality between the three output types. Separating signals enables more information to be carried for each signal. For example, in order for a composite cable to effectively multiplex video and luminance data, the signals must be combined in a way that they do not interfere with each other. Since the luminance signal is typically a much higher frequency than the color information, it is processed through a low pass filter prior to combination to ensure a lack of interference with the signals. This reduces the available bandwidth and harms image quality. Component and S-video signals are often sharper and more distinct than composite video, with greater reproduction of color fidelity.

It follows that standard definition camera systems can vary widely in output quality – sometimes even in implementations from the same manufacturer. Crawler manufacturers or end users will often use imagers from third parties – it is not uncommon to see Inuktun (now Eddyfi) imagers on Envirosight crawlers. While the imager may be capable of outputting component video, reserving three separate lines in the cable for video may increase the cost excessively, and the crawler manufacturer or systems integrator may reduce it to composite for transmission to the surface.

While analog television and 480i transmission has been removed from most consumer and business facilities for more than a decade, it is still commonly encountered in CCTV crawlers. Ruggedizing camera enclosures is a major engineering undertaking, and equipment manufacturers attempt to extend the life of products as much as possible. Further, in many pipeline inspections, the camera is only a few inches from the pipe wall, and optical zoom makes it possible to clearly focus on a feature. Thus, analog video has been "good enough" for sewer inspections. Further, analog video can be stored in small file sizes. When hundreds of miles of sewer lines are being regularly inspected, municipalities can quickly find that they don't have a server architecture capable of storing hundreds of terabytes of video data in a way that allows regular access. This means that end users have not yet pushed for contractors to upgrade to high definition systems.

It is important to note that many of these analog systems are not analog from end-to-end. Gone are the days of recording video to VHS tapes. Instead, there is a digitization step that occurs at some point in the inspection process. Often, video signal is fed into a control computer, digitized and recorded to a hard drive in a program like WinCan. That digitization process does not add quality – instead, it compresses and saves the analog video to a computer file – sometimes adding close-captioned textual overlays on top. Just because a video can be pulled up on a computer does not make it a high-definition, digital video. Beware also of unscrupulous users who will increase the resolution of the video file on the computer to create the illusion of higher data quality. A 1080p video file can be generated by stretching NTSC input – making all of the original lines wider, but not adding any information density into the video stream.

True high-definition digital systems are available, and the differences in image quality are just as stark as they were in the consumer space. Just as the transition from analog to digital television enabled viewers to see wrinkles in the faces of celebrities and shoddy repairs in the sets of popular television shows, so too can one's eyes be literally opened when viewing a high-definition sewer inspection video for the first time. While standard definition cameras can clearly show exposed aggregate in a concrete pipe, high definition cameras can give a clear picture of just how exposed that aggregate is – often times without zooming in closely or intentionally. This can make it much easier for inspection crews to notice problems, and some will attest that the higher image quality gives them a sense for the qualitative "feel" of a pipe (Figure 2.10).

Many new systems incorporate high definition imagers by default, and camera manufacturers are rolling out new products designed to facilitate easy upgrades. Users of the standard definition Inuktun SP90 can upgrade to the high definition SP120, though they will need to make some modifications to mounting brackets and video processing hardware. These digital imagers can output video in a variety of formats and do away with the interlacing used in the past. Instead, they rely on a technique called progressive scan, in which all lines are drawn sequentially. Thus, a 480p video signal can be twice as detailed (under some conditions) as a 480i signal. Most digital imagers can output video ranging from 480p to 1080p. Converters exist to downgrade digital video into signals compatible with older composite video technology – something that might be needed if a piecemeal upgrade is attempted. For example, an older screenwriter might rely

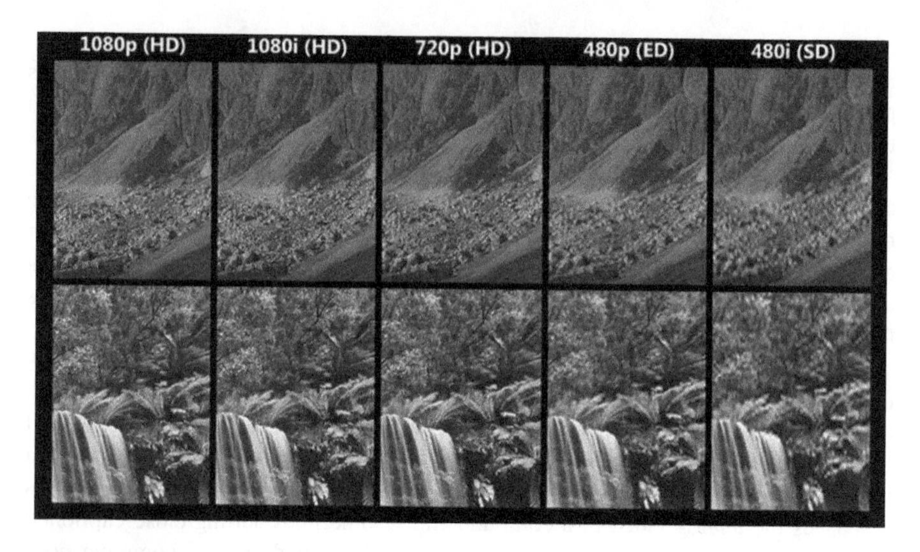

FIGURE 2.10 The jump in quality from SD (NTSC) to HD resolutions can be dramatic. Further, many digital imaging systems eliminate moire fringe and other analog artifacts. This image shows the same photo displayed in various resolutions from low-resolution SD to high-resolution HD. Image provided by Grayshi. Creative Commons: https://commons.wikimedia.org/wiki/File:HD_vs_SD_resolutions.png

on an analog signal, and video may need to be downgraded until that component can be replaced.

The biggest problem with digital video imagers is that they generate much more data. Digital video can be transmitted on a variety of cable types including component and HDMI – which either require more copper connectors in the tether, or some kind of digital conversion to fiber optics. This also means that digital video has a much higher data density than analog video. Actual file sizes depend significantly on other factors including bit rate and frames per second, but using statistics published by Avid Technologies – a maker of popular video editing software – some conclusions can be drawn. One minute of NTSC video occupies 1.22 GB of space, before compression. One minute of high quality 1080i video can occupy 8.68 GB. In a 45-minute sewer inspection, high definition video can generate tremendous amounts of data. As a result, digital video is almost always compressed in a lossy format, such as MPEG-2 or MPEG-4, in order to make file sizes manageable.

When crafting a specification for CCTV inspection, think carefully about what technology will be required to produce the information needed to manage a system. If most inspections are of 8–12″ lines, would high definition make a noticeable difference? Look at samples of video and ask yourself if it is worth the trade-off. Not only will high definition video increase your storage and management burden (talk with your IT staff about best practices), but it will also reduce competition in a bidding process. Not all CCTV inspection firms have high definition assets. Is it more important to get a lower price per foot, or more data for your lines?

The calculus will not be the same for every inspection program. While the question in the previous paragraph is deliberately left open-ended, there are many situations in which standard definition inspection of small pipelines would be perfectly acceptable. In large-diameter interceptors, the same assumptions may not apply. When inspecting a 120″ interceptor that has been expensively dewatered, it may be worthwhile to insist on the highest-quality data collection possible. Vendors capable of inspecting these large lines would be more specialized anyway, and perhaps it would be worthwhile to insist on one that keeps current with changing technology.

Regardless, in order to maximize competition, draft any specification to align with public video standards. Equipment should produce 480i (or better) video for a standard definition inspection, and 1080p for a high definition inspection. Some vendors may have produced custom cameras that output 1293 lines of resolution – to cite a hypothetical example. Don't tailor your spec to that resolution, and exclude other competent vendors with high-resolution video cameras. The example specification in Appendix A (p. 247) provides useful language to use as a guide. When in doubt, research video formats before making a commitment.

Fisheye cameras (sometimes referred to as side-scan cameras) work in a very different way and require a different set of requirements for complete analysis. Lenses on traditional cameras focus light from a subset of the area in front of the camera onto a digital imager, known as a charge-coupled device (or CCD). Higher quality lenses use expensive glass and precise optics to focus light more accurately and produce a sharper picture. Specialized lenses can ensure that more of the image stays in focus, by expanding the field of view. All work by bending light in a specified way to focus it on the imager.

Fisheye cameras or 360° cameras have lenses with a very different structure (similar to the eyes of many fish), and work in a way that is similar to virtual reality cameras. Virtual reality cameras have special lenses that bend light from a wide angle around the imager, capturing a ring of data that is effectively a panorama instead of a pinhole. This ring can extend to 140–180° around the lens, and even farther if expensive equipment is used. This generates a strange-looking video that is highly distorted, yet when that video is projected on the inside of a semicircle or cylinder, it regains its original sense of perspective. When a person is given equipment that lets them look around in that cylinder or sphere, they are given a realistic impression of being inside the environment. Expensive virtual reality systems can enhance this effect by stitching together imagery from a variety of cameras, enabling the viewer to smoothly look in all directions (Figure 2.11).

Fisheye cameras work in the same way, albeit in a less visually appealing environ ment. By mounting the same virtual reality-type lens onto a crawler, it is possible to perform an inspection that captures video data from the front and sides of the robot simultaneously. This means that robots equipped with fisheye lenses can simply drive down the pipe and return home, without stopping to PTZ at every defect or feature. Office staff can virtually drive through the projected video and PTZ from the comfort of a climate-controlled workplace – in theory generating more detailed inspection reports than crews out at low flow times (middle of the night) in the cold of winter.

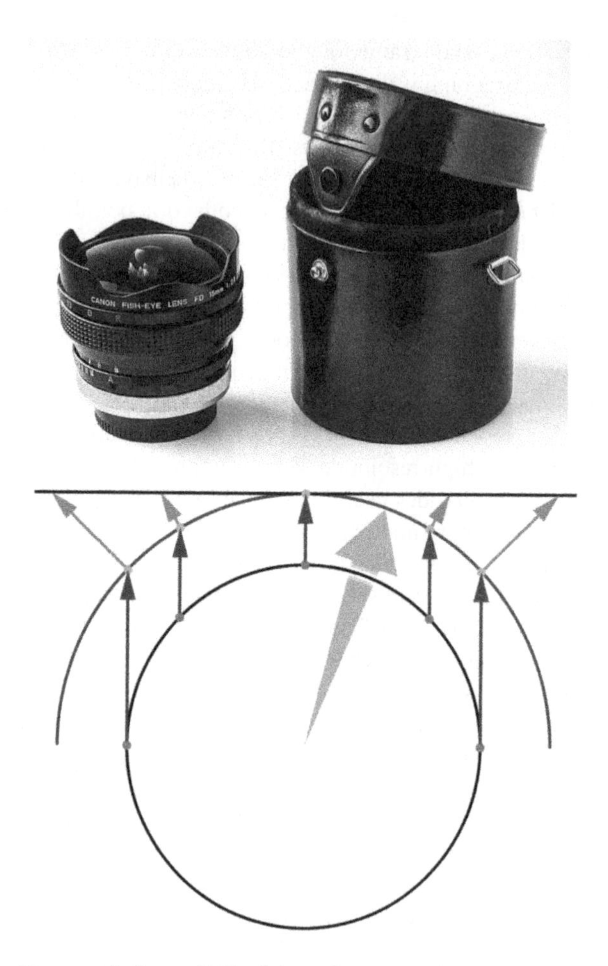

FIGURE 2.11 Commercially available fisheye lenses can be used to capture wide angle views, though they must project a three-dimensional environment onto the two dimensional CCD. This can cause distortion, especially at the edges, where fewer pixels are available to represent large amounts of data. Lens photo was taken by Franz van Duns and the projection diagram is from Peter Wieden. Creative Commons: https://commons.wikimedia.org/wiki/File:Canon_15mm_f2.8_SSC_lens_and_satchel_(_DSF3158).jpg, https://commons.wikimedia.org/wiki/File:Fisheye-view_projection.png

Some crawlers have more than one fisheye lens equipped to enable a coder or analyst to look in front of, beside, and behind the crawler at any moment in time. Many crawlers with 360° cameras take the technology one step farther by unrolling all of the acquired images and stitching them together – forming a map of the entire interior of the pipe. This enables analysts to look at the entire pipe at a glance, and quickly spot defects or areas of concern. Since this technology enables an easy look at the imagery to the side of a crawler, it is often referred to as side-scan technology. Both side-scan imagery and fisheye lenses use the same underlying techniques to generate this data, and can be considered variants of the same technology. However, not all systems that generate side-scan maps can generate virtual reality PTZ footage and vice versa.

Fisheye lenses offer the potential for significant improvements to the data collected during a PTZ inspection. However, since they operate in a very different technological manner, they must be analyzed according to a different set of criteria. First, fisheye lenses do not generate video that is akin to broadcast video, so terms like NTSC, 480p, and 1080p are not relevant. Instead, the resolution of a fisheye lens is driven by the underlying imager, much in the same way that a digital camera is. These imagers are often rated in the millions of pixels (or mega pixels) that they can capture per image. Thus, an 8MP camera can capture four times the pixels that a 2MP camera can.

Most vendors of fisheye cameras advertise the megapixel capability of the underlying CCD imager. This statistic is misleading, as most of these CCDs are square devices. Fisheye lenses project a circular image (when deconstructed in 2D) – meaning the image either does not fully cover the CCD or completely covers it and spills over the side. In the case of the former, some pixels on the imager go unused. In the case of the latter, some of the degrees in the available field of view captured by the lens are lost. Ergo, evaluating and comparing different fisheye lens systems is not as simple as comparing megapixels to megapixels.

Further, it can also be extremely difficult to compare fisheye systems to traditional CCTV, due to variations in the ways in which the lenses bend light onto the CCD imager. In a traditional lens, the degree to which light bends does not change significantly over the entire surface of the lens. For example, even at the edge of a traditional lens, light may bend a few degrees. However, at the edges of a fisheye lens, light rays bend 70–90° as they travel through the lens and make contact with the imager. These outermost degrees of a fisheye lens are the most important, as this is the area that corresponds to a direct view of the pipe wall. Yet, comparatively, this information is spread across fewer pixels than is information at the center of the lens.

One way to visualize this is to consider the CCD with a series of concentric rings overlaid atop it – similar to a target. Assume all the rings are the same width. The innermost ring captures data that is almost directly in front of the robot – a span of maybe 1–2°. The next ring is capturing a slightly broader area, perhaps 3–4°. Since more degrees are being projected onto a band of pixels that is the same width, some detail will be lost – likely not much, as the resolution is close to the adjacent ring. However, by the time we get to the last ring, the same width of pixels is responsible for imaging a 10–15° swath of data, suggesting a 10x reduction in information density compared to the innermost ring.

CCDs are not actually broken down into concentric rings, but pixels are discrete units and cannot be subdivided. While there will be no banding on fisheye images, the image degradation from the center to the periphery will still be significant, albeit gradual in its progression. Thus, an image that looks crystal clear down the center of a fisheye lens can look blurry, pixilated, and distorted when one pans to look at the edge of a pipe wall.

This can be surprising, because in terms of raw megapixels, fisheye lenses are capturing far more information than their conventional counterparts. A conventional 1080p camera has a resolution of 2.1 megapixels, and commercially available fisheye cameras can have resolutions of 8 megapixels or more. Yet, a virtual PTZ inspection on the fisheye camera is often a lackluster experience. The reason is similar to digital zooming. When an analog PTZ camera is positioned to look at a defect, the full range of 2.1 megapixels is focused on that defect and is used to represent it with as much detail as possible. With a fisheye virtual PTZ, some subset of the available 8 megapixels is expanded to fill the full screen in the location of the defect (just as digital zooming involves expanding a subset of an image). This, coupled with the fact that images of the pipe wall are confined to the outer edges of the CCD, often leads to a defect image of less than a megapixel – a disappointing conclusion from this futuristic technology (Figure 2.12).

This is not to say that fisheye cameras cannot produce breathtaking images – they can – but they often require high-quality lenses and extremely large CCDs to do so. Commercially available GoPro cameras often use more than 16 megapixels to generate acceptable 360° video, and similar or larger resolutions will likely be required for sewer inspections, due to the necessity of information at the periphery.

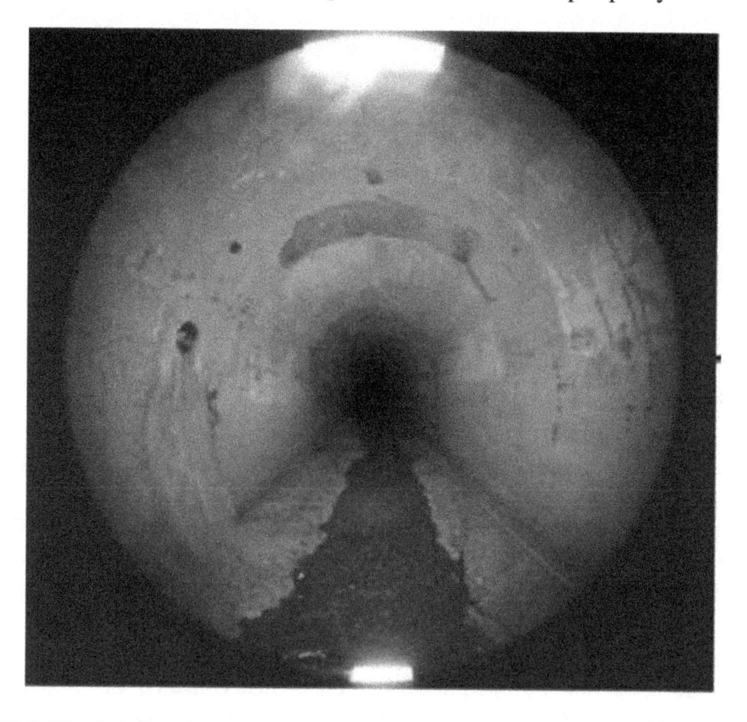

FIGURE 2.12 A fisheye image taken of the interior of a pipeline. Notice the distortion at the edges. Software algorithms can "dewarp" these images and correct these artifacts. For the appearance of depth, the images can be projected onto a 3D shape and viewed in virtual reality. Image taken from Guo et al., *J Comput Civ Eng.* STM (ASCE): https://ascelibrary.org/doi/pdf/10.1061/%28ASCE%290887-3801%282009%2923%3A3%28160%29

Finally, users new to fisheye camera systems are often surprised to find that the default inspection output format for these systems consists of a low framerate video, where obvious jumps from frame to frame occur on the order of 2–5 frames per second. For those of us who are accustomed to 30 fps video on our home televisions (and 24 fps at the cinema), the result can be jarring. Unfortunately, this stems from the size of the data that is being collected. Recall that 1 minute of 1080i footage can generate almost 9 gigabytes of data. An 8MP fisheye camera would generate four times that amount or 36 gigabytes per minute. A 30-minute inspection could overwhelm a 1 TB hard drive. The solution is to reduce the frame rate and capture images at a much lower frequency. So long as the robot does not move too quickly, this can be acceptable and still produce a high-quality side-scan image, even if the video can be disorienting.

Why then would one even consider adopting fisheye technology? There are three main arguments that proponents adhere to: first, as robots move toward autonomy, manually panning, tilting, and zooming to focus on each defect will not be possible, as there is too high a risk of missing something. Instead, autonomous systems will likely navigate through a pipeline in one linear motion, capturing as much data as possible, passively. This way, the crawler need only focus on traversal, knowing that enough data will be captured regardless of activity. Subsequent analysis can take place under the guidance or supervision of a trained human before information is discarded.

Second, while the process of recording and classifying defects will be described more in the next chapter, there is some evidence to suggest that the loss of resolution can be offset by a higher quality analysis that comes from separating the act of driving the crawler and actually looking for defects. While no one is suggesting an inherent difference between field crews and office workers, the conditions that result in a meticulous inspection are often very different from those present in the course of a typical sewer inspection.

Many sewer inspections take place under challenging conditions – some manhole access points are located in the middle of busy roads, necessitating expensive traffic control services. Other sewer lines typically have so much flow that they can only be inspected during low flow times, when most of the population is sleeping – think 12–3am. While it is difficult to conduct sewer inspections during a torrential downpour, inspections do occur during times of freezing and sweltering temperatures. Working under these challenging conditions and with a long list of work orders ahead of them, it can be difficult for many crews to curiously investigate small defects.

Instead, fisheye technology means that crews need only focus on getting the robot through a pipeline successfully, knowing that if that occurs, enough data will be gathered for a successful virtual PTZ or side-scan inspection. Office workers are then tasked with differentiating roots from cracks from shadows, and can do so from a climate-controlled environment. Further, because these office workers can specialize in analyzing footage, they are able to have much higher throughputs and devote additional time to QA. While not every sewer analyst is perfect, fisheye technology sets up a division of labor that lends itself to higher-quality inspections (Figure 2.13).

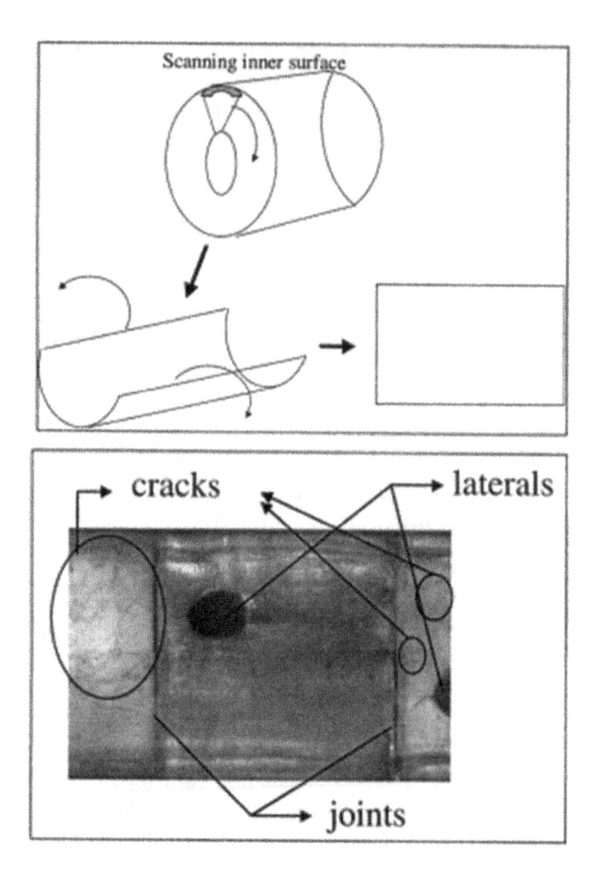

FIGURE 2.13 Side-scan sewer inspection data can be "unrolled" to produce a single image consisting of the entire visible portion of the pipe wall. The method of unrolling is shown schematically in the top image, and a segment of unrolled pipe is shown in the bottom. Image taken from Chae et al., *J Comput Civ Eng.* STM (ASCE): https://ascelibrary.org/doi/pdf/10.1061/%28ASCE%290887-3801%282001%2915%3A1%284%29

Finally, fisheye technology enables faster inspections. If a municipality is facing a consent order deadline, or wants to inspect all of its lines in a short amount of time, it may have no choice but to adopt fisheye technology. Many fisheye devices are floating or autonomous, meaning they can cover long distances, or be deployed in closed manholes, minimizing traffic control and confined space entries. Further, they can cover much longer distances at a time. RedZone Robotics estimates that one person can successfully manage a squad of 4 Solo robots, and under some conditions, that sole individual can inspect 8,000–9,000′ of pipelines per day. Conventional teleoperated CCTV crews can inspect 1,000–1,200′ on average – efficient use of fisheye technology can enable a significant increase in productivity!

How can fisheye technology be integrated into a conventional specification? First, it is important to differentiate between different degree measurements. While some refer to fisheye lenses as 360° lenses, that term can be misleading. While fisheye lenses do capture data in a 360° circle around the center of the lens, the side-to-side field of view ranges from 140° to 180°. A device with a 180° field of view will be able to look at a defect face on. A device with a 140° field of view will only be able to image defects with a slight forward skew. Both mean that it will not be possible to look up a lateral unless a second camera is employed.

Thus, rather than specifying degrees and pixels, sometimes functional specifications are more appropriate. Specifying a 1080p PTZ camera that is capable of looking up a lateral or a fisheye system of equivalent or better functionality will enable vendors of both inspection technologies to bid for a contract. Specifying that both systems must be capable of detecting a crack on the wall of a 60″ sewer provides another functional standard. A conventional camera system will easily be able to resolve this feature through PTZ and optical zoom – if a fisheye system can meet this standard, the vendor should be able to provide sample videos or data files that prove it.

Most importantly: beware the bait and switch. Deployment costs of many fisheye systems are much lower than conventional CCTV crawlers. Don't let a vendor bid a job with a conventional system, and then switch to a fisheye system without notification. Since fisheye systems have higher throughput, they should come at a lower price per foot, especially considering the data quality issues they may introduce. Swapping inspection devices like that is significant enough to require a written, documented change order, and language to that effect should be in any contract.

Finally, if you are considering allowing fisheye technology in an upcoming contract, familiarize yourself with what the actual deliverables look like. Virtual PTZ inspections are an incredible tool, but will you need to install a custom viewer to perform them? Can you do so on your municipality-controlled computer? Is it possible to generate virtual inspections in standard video formats so that this data can be integrated with data from other vendors? Watch some actual inspection footage and get accustomed to the actual frame rates in use. Can any side-scan images be loaded into an asset management tool, or will they need to be broken up? Uniform answers to these questions have not yet been developed, so take the time to understand a technology carefully before committing to it – don't expect standard CCTV video by default.

Many CCTV specifications emphasize the ability of a camera to look up a lateral – in order to observe flow activity, root intrusion, and other features of note. Many modern crawlers can go beyond that with dedicated lateral launchers. These units involve high-strength springs and integrated push camera functionality in the head of the crawler. When one of these units looks up a lateral, it is possible to detach the imager, and actually push it up into the connecting line to look for features. In terms of efficiency, this feature can dramatically increase productivity. A single mainline inspection can accomplish hundreds of lateral inspections simultaneously – a major concern when looking for crossbores or other time-critical tasks.

Most CCTV systems are tethered in order to enable teleoperation. This means units have a long cable with multiple strands of copper connectors and/or fiber optics within to enable the transmission of power and data to the unit. Copper systems experience attenuation with length and tend to max out at 1,000–2,000′. Fiber-optic systems can be much longer – 7000′ reels are not unheard of. However, fiber optic cables are much more expensive and difficult to terminate. Steel reinforcement integrated into these tethers enables their usage as load-bearing tools to retrieve stuck crawlers (though only to a point).

These tethers are fed through a rotary encoder that measures the chainage of the crawler during the inspection. Several errors can occur during this process – first, make sure that crews communicate so that driver and reel operator work together to maintain tension in the cable. If the robot stops but tether continues to be fed into the manhole, the distance the robot travels will appear to increase, even though the robot is not moving. Similarly, before starting an inspection, follow a consistent process to align the 0′ mark with the interface between pipe and manhole. When digging to repair defects, errors of "just a few feet" can be extremely costly.

Finally, be aware of the fact that tethers tend to stretch during a deployment – physics dictates that robots will use more force to pull 1000′ of tether at the end of an inspection than they will near the beginning. This stretching means that locations and defects identified at the end of an inspection may not be as accurate as those identified in the beginning. Always confirm manhole distances with GIS data or GPS measurements, and recognize this discrepancy. If an inspection contractor claims a 1% accuracy in chainage, a 10′ discrepancy in a 1,000′ inspection should not be alarming. Similarly, that 1% accuracy when applied to a 7,000′ inspection could mean that defects could be off by up to 70′ from their observed location – something that should be considered in detail before engaging in maintenance or repair operations – especially those that will result in surface disturbances.

Encoders can lose calibration over time, so insist that vendors show recent calibration records for tether systems. If a vendor is claiming 1% accuracy, they should be able to show tests within the last six months showing <10′ of discrepancy in a 1,000′ tether. Make sure that calibration data is from the actual crawlers and payout encoders that will be used on your inspection. All of these devices should have easily verifiable serial numbers – if they don't, run away from that vendor! Unserialized equipment will make it difficult or impossible to track down and correct errors if they are discovered after inspections have completed.

INSPECTION CONDITIONS: CLEAN TO INSPECT OR INSPECT TO CLEAN?

When launching a comprehensive CCTV inspection program, some thought should be given to the goals of the effort when deciding where in the overall process inspection should occur. Preparation and cleaning can have a significant impact on the quality of data generated from an inspection program. However, this prework comes at a cost that may or may not be worth it for a municipality (Figure 2.14).

When inspecting water lines, there is really no debate. In order for a crawler to inspect a water line, the line must be dewatered. As water lines are pressurized, they often encounter changes in grade, making inspection a difficult endeavor for crawlers that were designed to traverse gravity sewers. Regardless, it is extremely difficult

FIGURE 2.14 In some cases, pipelines truly must be cleaned prior to inspection, but the data does not always support this conventional wisdom. Why structure contracts as clean to inspect, when some pipelines may not need to be cleaned at all? Image provided by Hammelmann Oelde. Creative Commons: https://commons.wikimedia.org/wiki/File:Pipe_cleaning.jpg

to remove tuberculation from a water line prior to inspection – thus, most are televised in their recently dewatered conditions.

With sewer lines there are additional choices to make. Many sewer lines contain grit and debris in the invert of the channel. Fats, oils, and grease deposits can be stuck to the sides of the pipe, and roots can impinge on available space for a crawler. Encountering these debris often leads to the abandonment of a survey – if the crawler cannot get through them or the lens becomes completely occluded, it is impossible to complete an inspection. As a result, the de facto standard for smaller diameter lines has been to structure programs in the form of clean to inspect – all lines are flushed with water, or cleaned in some way, prior to inspection.

Perhaps there is no harm in this practice: after all, regular flushing of small diameter lines ensures that regular scouring takes place, and prevents the likelihood of harmful debris buildup. It also increases the chance that inspections will proceed to completion and generate data about the entire interior of the pipe wall – no cracks will be covered by grease deposits, for example.

Yet, it is important to note that while this practice is designed to maximize the generation of useful information, precleaning washes away valuable data about debris buildup in addition to the debris themselves. If all lines are cleaned in advance, it can be difficult to recognize trends about debris buildup. How many lines were cleaned in which there was no real need to do so? Further, examining the debris present in a pipeline can reveal information about underlying problems – fresh-looking soil near a tap could be an indicator of a break in the lateral causing significant inflow. Deposits at joints could indicate offsets that are difficult to see from the crawler's default inspection perspective. Rather than adopt a blanket cleaning schedule, could a differentiated policy save the municipality money in which the dirty lines are cleaned more frequently than the clean ones?

Unfortunately, the structure of existing contracts works against this data generation model. Contractors are typically only paid for complete inspections – instances in which the crawler makes it from the start manhole to the finish manhole. This means that they will advocate for precleaning as well, as it enables them to get

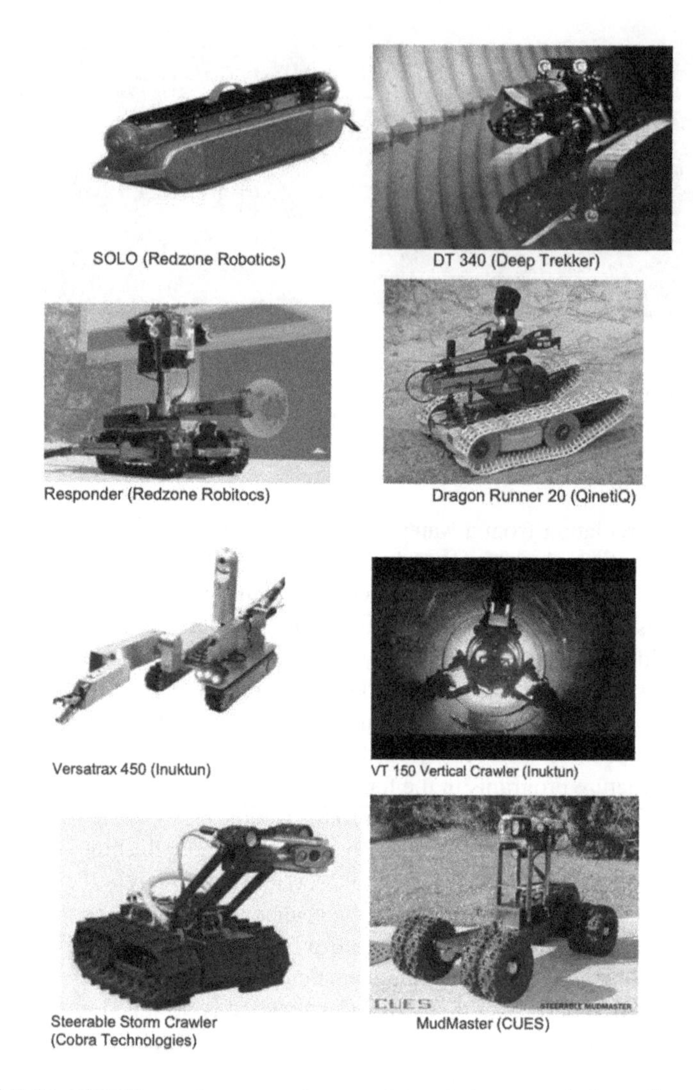

FIGURE 2.15 CCTV crawlers come in a variety of shapes and sizes. Some, such as RedZone's 700 lb Responder, are intended for large interceptor inspections. Others are intended for much smaller lines. Note the variety of propulsion systems and imaging technologies. Public Domain: https://rosap.ntl.bts.gov/view/dot/34913/dot_34913_DS1.pdf

through more lines. Further, the fact that many of them have jetters and vac trucks in their fleet provides ample opportunity for additional profit. Instead, municipal decision makers should consider adopting a contract structure that pays for feet of inspected pipe – regardless of whether or not they result in full televising of a line. If an inspection fails due to debris buildup, that is valuable information to have. The contractor should be paid for providing it, and the municipality has the option for remediation and to perform a postcleaning inspection after the fact.

While this may seem more costly – performing two inspections, rather than one – when the costs of systematic cleaning are taken into account, the savings may not be as dramatic. In some cases, municipal customers who adopted an inspect-to-clean approach found that only 25–30% of their lines needed to be cleaned at all. A full 75% of the system was fine. That savings in recurring operation and maintenance costs could be deployed toward other priorities or even new capital projects if budget flexibility allows.

Strangely, this attitude persists for the more difficult–to-clean interceptor systems as well. Cleaning these larger diameter lines can involve much higher costs due to the difficulty of access and specialized equipment required. While the EPA requires utilities to make best efforts to maximize system capacity, that does not mean that lines with minimal debris and sediment should be cleaned regardless of need.

Many contractors will not like this approach, as their equipment cannot handle traversing lines that have a moderate amount of debris – think 6–12″ in a 144″ interceptor – and insufficient efforts are unlikely to lead to payment in full. Floating systems or rugged devices like the Responder, by RedZone Robotics, can navigate through these lines even under *in situ* conditions. However, deploying the 700lb, hydraulically powered Responder robot comes at a higher price point than attempting the same task with a commodity crawler (Figure 2.15).

3 Classifying and Analyzing Visual Inspection Data

The same quality that makes visual inspection data so pervasive contributes to the difficulty inherent in its analysis. Collections of videos and images are qualitative and lend themselves to subjective interpretation. What constitutes a large crack? Does this surface suffer from moderate wall loss? Are these roots fine or medium in consistency? While some defects can be universally agreed upon – few will argue that roots are anything other than roots – reasonable people can disagree about the severity and adjectives used to describe defects in sewer systems.

Thus, for as long as visual sewer inspections have existed, so too have data classification systems for standardizing and interpreting inspection data. In North America, the most common standard for this is the Pipeline Assessment and Certification Program (PACP) created by the National Association for Sewer Service Companies (NASSCO). PACP is derived from a slightly different European standard – the Manual of Sewer Condition Classification (MSCC) established by the Water Resources Centre (WRc) – which is still widely used in Europe, the United Kingdom, and Canada. Other countries have their own inspection standards which are similar in structure, yet different in nuance.

ROOTS OF CONDITION ASSESSMENT: WRC'S MSCC

Before NASSCO created the PACP program, different municipalities had their own ways of coding and classifying data. Some kept detailed handwritten or typed logs, which accompanied cabinets full of VHS tapes. Others developed their own forms or abbreviations to standardize notes among sewer inspection crews. For example – a municipality might differentiate between small, medium, and large root balls with the abbreviations RTS, RTM, and RTL, respectively. Any crew encountering roots could then just use standardized shorthand rather than trying to qualitatively describe the roots to the best of their abilities.

Unfortunately, different municipalities developed the need for these standardized systems at different times, and in relative isolation. Thus, throughout the end of the 20th century, different standards for sewer inspection began to appear in cities and regions around the United States. While this was not an issue for in-house municipal crews, third parties had to conform to the standard of the municipality in which they were working. Failure to adjust led to errors, miscodes, and a poor quality of work when transitioning from municipality to municipality. Further, as consulting engineers began professionalizing the management of wastewater collection systems, this lack of standardization made it difficult to draw conclusions from data generated by multiple municipalities.

The United Kingdom was the first to rectify this issue. In 1980, the WRc published the first edition of the MSCC in order to provide standardized guidance to inspection crews. The work was so successful that the codes and format specified in the MSCC have gone on to become British Standards, officially used throughout the United Kingdom. Further evolution of the MSCC has aligned it with European Union standards in its most current revisions. The MSCC is often paired with the Model Contract Document for Sewer Condition Inspection, establishing a standardized set of equipment and best practices for any sewer inspection performed within the UK.

The MSCC came to standardize buried pipeline inspections in several key ways: first, a detailed header was filled out for every inspection. This header contained information about the name of the person completing the inspection, the location of the pipeline, GPS coordinates, notes about the construction material, depth measurements, weather information, and more. This enabled inspections to be queried against a range of variables, and made it possible to refine, localize and isolate issues. Simply recording the name of the inspector is a key example – if an inspector was poorly trained or made a systemic series of errors, it is easy to find all inspections that could have been affected. Prior to MSCC, the name of the surveyor was not always permanently attached to a survey.

Construction material is also significant. While the materials used in sewer construction tend to be obvious to those in the field (bricks are seldom mistaken for concrete), including it in the inspection record can prevent subsequent errors. If an engineer designing a rehabilitation program or lining insert glances at a survey and notes that it was collected in a concrete pipe, but he is preparing a liner for a brick system, it is immediately obvious that something is amiss. Perhaps the as-built drawings are incorrect, or perhaps different lines were investigated. In either case, quickly noting the material in a searchable format can prevent costly mistakes down the road.

Some may fail to appreciate the inclusion of weather information. While it is true that individuals reviewing an inspection record do not really care if the weather that day was sunny or partly cloudy, the presence of precipitation can have a major impact on observed conditions in combined (and sanitary) sewer systems. If flow levels are in excess of 75% of the diameter of a combined interceptor pipe, and it is a dry day, it is apparent that the system will likely experience severe capacity issues. If it has the same flow levels occur during unprecedented rains, it may be much less troubling to the individuals responsible for planning and maintenance.

MSCC also worked to describe defects in a standard, codified way. The location of defects in a pipe was written in terms of an inspection direction, chainage, and clock position. The combination of inspection direction and chainage enables localization of a defect at any point along a vector representing a sewer line. For example, a crack located 180 m upstream of the start manhole can be plotted on a geographic information system (GIS) in order to easily evaluate repair and rehabilitation potential. The clock position establishes where in the pipe's cross section the defect is located. Defects at 12 o'clock are located at the crown, 3 o'clock on the right side of the spring line, 6 o'clock in the invert, and 9 o'clock at the left side of the spring line. All clock positions are relative to inspection direction.

When repairs and defect remediation are thought of in terms of digging or lining, the clock position can seem insignificant. However, if illegal taps are being noted, the clock position combined with the chainage can not only identify where on a given street the offender is, but also on which side of the road. Without the clock position, it could be difficult to translate a tap to an address and take appropriate action. Clock positions also eliminate the vague and confusing descriptive references individuals would come up with – a crack located 60% above the flow line on the right side, versus a crack located at 2 o'clock. While the former may be well meaning, changing flow levels can make even copious notes useless under different conditions. The latter is clear, concise, and will remain true so long as the pipe remains in its current location.

Clock positions can also describe a range of locations within the pipe segment. For example, if a defect wanders around the crown of the pipe, it could be said to occur between 11 o'clock and 1 o'clock, or 1101. Hydrogen sulfide corrosion occurring in the upper half of a pipe segment could be noted as spanning from 9 o'clock to 3 o'clock or 0903. The order of the clock positions matters, as the defect is assumed to travel between the noted locations in a clockwise direction. 0903 refers to the top half of the pipe, whereas 0309 refers to the bottom half. 1101 represents a narrow range near the crown, whereas 0111 indicates the bulk of the visible pipe segment, including the invert.

Each defect is represented by a short alphabetic code, the makeup of which is related to the words describing a defect. For example, cracks are represented with the letter C, infiltration with the letter I, missing mortar with the letters MM and displaced bricks with the letters DB. Additional letters can describe the defect in more detail; for example, a crack running down the length of the pipe is a longitudinal crack, represented by CL. A crack moving around the circumference of a pipe is a circumferential crack, or CC (Figure 3.1). Some codes have no additional lettrage – missing mortar will always be MM – no distinction is made between the type, depth, or pattern of the missing mortar. The only item of note is that it is indeed missing.

Not all defects of a given type are created equal – some defects may be more or less serious depending on where they occur. For example, root intrusion at the point at which two pipe segments come together (a joint) is undesirable, but understandable – the nutrient-rich sewage provides a tempting source of sustenance for trees and other plants. That same root intrusion in the middle of a pipe segment would be bizarre and suggestive of other failures that might be hidden under the roots themselves. To account for this, the joint modifier can be applied to almost any code to denote whether or not it occurs in proximity (0.2 m) of a joint between pipe segments. This can be a critical piece of information for engineering or rehabilitation decisions.

Defects and observations are also modified with a summary of quantification. This can take the form of dimensions, percentage, or band. Dimensions represent one or two measurements corresponding to a physical size, rounded to the nearest 5 mm. This could be used when the pipe changes diameter, for example. The new diameter of the pipe would be entered to the nearest 5 mm – either based on as-built drawing information or a visual estimate. Percentage refers to percentage of either

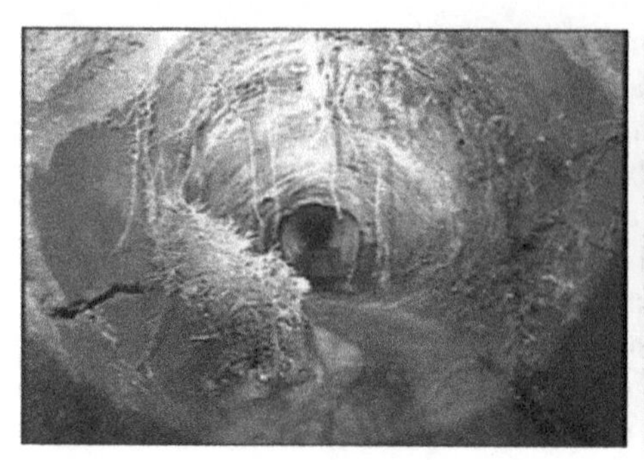

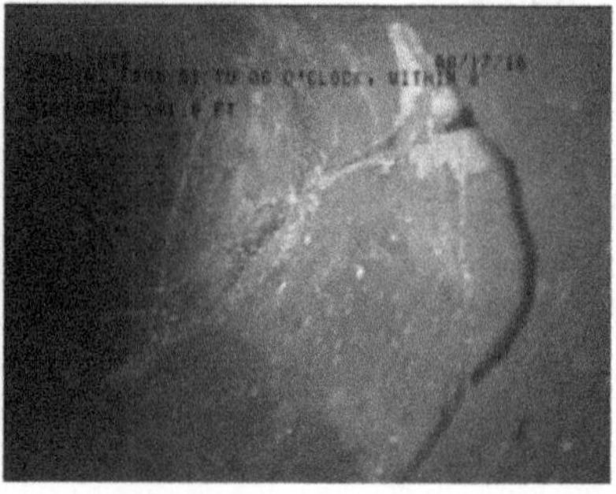

FIGURE 3.1 Defects can aggregate, as one defect can cause others. A broken, cracked pipe can provide a point of entry for roots, which can further open cracks into fractures. Thus, the distinction between O&M and structural defects is interesting, but both types require corrective action. Public Domain: https://nepis.epa.gov/Exe/ZyPDF.cgi/P100C8E0.PDF?Dockey=P100C8E0.PDF

the diameter of the pipe or cross-sectional area. Water levels are entered in terms of the percentage of the height of the pipe – a half full flow would be coded as 50%. Deposits attached to the side of a pipe would be coded in terms of the percent of cross-sectional area reduction. Both are coded to the nearest 5%. Knowledge of the underlying code syntax is necessary to understand which type of percentage is being used.

Finally, band is used when percentage or dimensional quantification is not logical. For example, it can be difficult to visually estimate the amount by which a joint between two pipe segments is offset. Perspective, lighting, and angle can make joints appear to be much more widely offset than they really are. Open joints can be classified with the M band if they are medium – offset by 1–1.5x the pipe wall thickness – or L if they are large – offset by more than 1.5x the pipe wall thickness. Joints that are open by more than 20% of the diameter of the pipe should have the opening estimated in mm. Similar bands can be used to describe displaced joints and other defects as well (Figure 3.2).

Sometimes a defect is not an isolated, discrete observation: instead it refers to something that extends along the length of a pipe segment – such as a crack, abrasion, or even root growth. If such a defect is encountered, it can be tagged with the continuous modifier, rather than being noted by the observer at regular intervals. These continuous defects are "opened" at the starting chainage and "closed" at the ending chainage. If a defect changes in size or location, one continuous defect must be ended and a new one begun. For example, if a pipe has a longitudinal crack at 9 o'clock for the first 50 m, and then that crack disappears and a new one emerges from 50 to 100 m, each must be coded as its own continuous defect.

Continuous defects can refer to two different types of continuous defects. Truly continuous defects are just as they sound – defects that continue down the pipe for a significant amount of chainage. For example, fine roots running down the pipe for six meters would be an example of a truly continuous defect. However, continuous

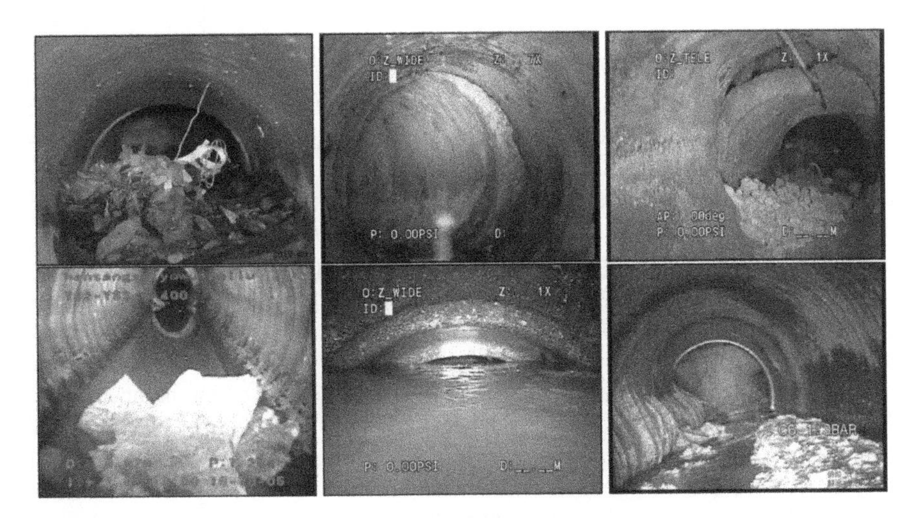

FIGURE 3.2 Modifiers are critical for accurately describing defects. The term "debris" may conjure images of a silty sediment, but as these photos indicate can encompass a variety of materials and quantities. Notice that some of these debris would be considered deposits, while others would be classified as obstacles. Image taken from Li et al., *Automation in Construction*. STM (Elsevier): https://www.sciencedirect.com/science/article/pii/S0926580518306174

defects can also refer to repeating continuous defects. If every joint is open by a medium amount, one code can be entered and allowed to stay open as long as the defect repeats regularly. Similarly, if roots intrude at every joint, one continuous code can describe this. If the continuous modifier is used for repeating point defects, those defects must occur at almost every joint that is encountered. The two types of continuous defects can be distinguished by the presence of the "joint" modifier in the underlying code. If the joint modifier is present, the defect is repeating. If the joint modifier is absent, the defect is truly continuous.

MSCC was created to standardize inspections in an era before computer systems. In that time, operators would complete a televised (or manned entry) inspection, and note defects on a paper form, referencing certain times on a VHS tape or still images that were produced. Because of this structure and format, it is a very rigid system, conserving space and promoting efficiency whenever possible. In today's world, the vast majority of inspectors will be performing their inspections using software packages like WinCan or IT Pipes, which replicate the contents of the form in a graphical user interface. Some inspectors may feel that MSCC forces them to choose between two options – neither of which truly represents the condition they are observing because MSCC views it as a binary choice (dating back to the paper form era). One should always remember the remarks field can be used to enter freeform text to describe something that doesn't fit into the classification system.

Further, while it is important to describe every defect accurately, remember that CCTV inspections are inherently qualitative and require judgment calls. Two well-trained observers can view two sections of reinforced concrete pipe and disagree about whether the roots that are intruding are fine or medium. Similarly, they can go back and forth about whether or not grease deposits represent a 5% or 10% reduction in cross-sectional area. Anyone making a rehabilitation decision based on a single code will certainly look at the video of the section in question and confirm the coding as it is employed. Decisions about replacing entire pipes typically rely on multiple codes to make an overall conclusion about a pipeline – thus while attention to detail is essential, operators should do their best, feel confident in their assessments, and not become paralyzed by the nuances of every last code.

Newer versions of MSCC continue to add and remove codes to make things clearer and more standardized. Artifacts of the analog era are gradually removed, and new digital features and codes in line with the rest of Europe continue to be added. Each new revision of the manual is indicated with a version number. Thus, MSCCv5 supersedes MSCCv4. Remember, when specifying a contract for CCTV inspection work familiarize yourself with changes between revisions. Don't just assume that you'll understand all of the new codes in v5 because you were trained in v4 – take the time to take a refresher course and learn about the new features. Remember, contractors will typically upgrade their software as soon as a contract requires them to do so, so it is better to stay ahead of the market in terms of familiarity with new standards. While newer versions of inspection software can often code in compliance with older MSCC revisions, inspectors will begin to lose practice with the older systems, so staying current can improve data quality.

This can sometimes be a tall order – years of pipe inspections may have been classified and categorized under older revisions of MSCC. That doesn't mean that a well-managed utility should stay frozen in time. Inspections that were coded according to MSCCv3 could date back to 1993–94, and the lines they correspond to are likely in need of reinspection. Some clever database work can help a system manager draw conclusions between old and new inspection data, and the presence of older formats should not be used as an excuse to delay or slow down modernization.

NASSCO PACP

While the MSCC continues to be revised and sees active use in Europe and in some parts of Canada, the United States has largely adopted the Pipeline Assessment and Certification Program, developed by the National Association of Sewer Service Companies. PACP started as a modification of MSCC for American audiences, but quickly evolved into its own coding system. Many codes in PACP can be traced directly to their ancestors in the MSCC, and even use the same images to illustrate the underlying defect or observation. However, in many cases, the nomenclature and abbreviations are completely different (Figure 3.3).

Some differences between the two systems are due to slight variations in how terms are described in the two countries. Others stem from timing: the MSCC was created in the 1980s and evolved to its current form through multiple iterations. The first PACP training course took place in 2002, and the system was developed in the years leading up to that point. PACP was created for a digital world. While it

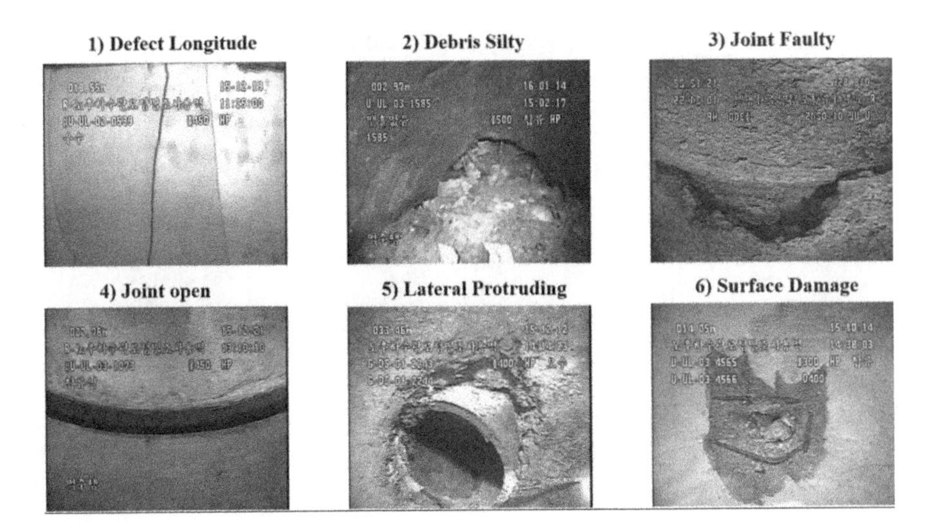

FIGURE 3.3 Note the Korean text on the CCTV overlay – while many regions have their own standards for classifying sewer defects, many common themes arise, even if the vocabulary is slightly different. Image taken from Hassan et al., *Automation in Construction*. STM (Elsevier): https://www.sciencedirect.com/science/article/pii/S0926580518305892

still relies on a detailed header and defect codes that could be written on a sheet if necessary, PACP data was designed to be captured on a digital system from the very beginning (in terms of codes, not video data). It wasn't until the late 2000s that compressed digital video could be stored on a computer's hard drive, rather than an external VHS tape or DVD disc.

Thus, where MSCC refers to an increase in surface roughness with the code SW, PACP refers to it as SRI. MSCC refers to open joints with the code OJ, while PACP refers to offset joints with the code JO (maybe due to a certain high-profile case in the United States in the 1990s?). Both use the same systems of chainage, clock position, joint codes, and continuous defects. PACP, however, eschews the confusing band concept, and ties modifiers into each individual code. For example, JO can be coded as JOM or JOL – the medium and large modifiers are simply optional parts of the code, rather than part of a separate classification for banding.

Both PACP and MSCC break defect codes down into two broad categories – structural defects, and operation and maintenance (O&M) defects (additional categories for observations – going beyond just defects – include construction and miscellaneous). Structural defects refer to issues with the underlying pipe itself that could lead to structural failure – things like cracks, holes, or major deformation. O&M defects are defects that could lead to blockages and contribute to eventual damage, but could be addressed with appropriate remediation and corrective action short of repair or replacement (Figure 3.4). For example, roots can be treated with a foam or cutter, and do not necessitate the replacement of a pipe segment. They can, however, cause future failure of the pipe if they are not addressed. Similarly, grease deposits reduce flow volumes, but can be cleaned off – albeit with difficulty.

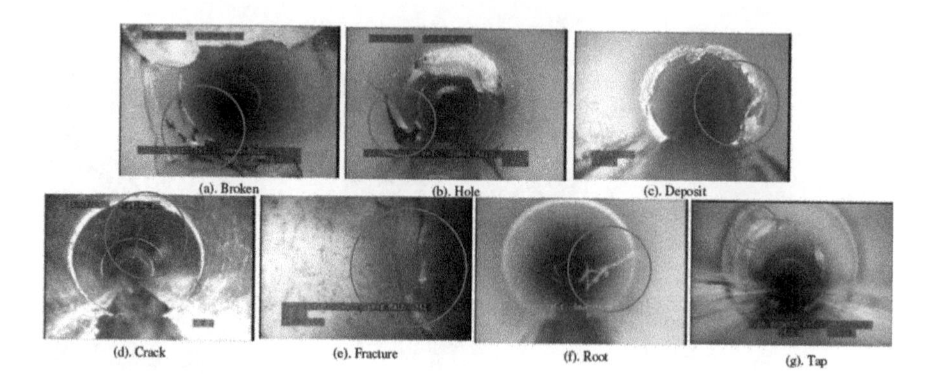

FIGURE 3.4 While some find it frustrating, CCTV inspections coded to standard note both defects and construction features. Lines with hundreds of taps can take hours to complete, whereas lines with few taps may be completed in a fraction of the time. STM (Elsevier): https://www.sciencedirect.com/science/article/pii/S0926580519307411

Many coding systems attempt to leverage this categorization by including algorithms to score or "grade" both defects and pipelines. In PACP, defects are assigned a numerical score between 1 and 5 as an indication of severity. Defects with a score of 1 are considered minor, 2 are minor to moderate, 3 are moderate, 4 are significant, and 5 are considered most significant. A single crack in a pipeline would be assigned a grade of 2, whereas a crack that opened enough to expose the interior would be upgraded to a fracture and would be assigned a grade of 3. The detail surrounding a defect can also impact the grade that is assigned. A section of broken pipeline would have a grade of 4, but if the break is so serious that the pipe surroundings are visible, then it is upgraded to a 5. It does not matter if the surroundings constitute soil or a void – it still becomes a most significant defect. In some cases, there are no modifiers: a section of collapsed pipe will always be a 5.

How and where a pipe is used will also impact the grading that is assigned to defects. Lines used for sewage, stormwater, or even fresh water that are in normal service receive one set of grades. Lines that are implemented within dams and levees receive a different set of grades that are usually more serious and tied to the dramatic consequences that could occur in the event of pipe failure. For example, small joint offsets do not even receive a grade in normal contexts, yet they are given a grade of 4 in dams and levees – presumably because any erosion through the joint could quickly compromise the structural integrity of the overall structure. Medium roots coming in through a joint are given a score of 3 in normal pipelines, but are upgraded to a 4 in dams and levees. Sometimes scores become less severe: ball barrel roots are considered a 5 in sewer/stormwater pipes, but only a 4 in dams and levees.

This categorization and labeling makes it easier for individuals charged with rehabilitation planning to triage and prioritize work (Figure 3.5). Computer software makes it possible to quickly sort through and see all lines with defects rated 4 or higher, in order to get a rough estimate of which parts of the system will require immediate attention. Inspectors and data analysts should use caution: it is not unheard of for a representative from a city to push back on codes when they don't

Code	Description	Code	Description
B	Broken	HSV	Hole soil visible
BSV	Broken soil visible	HVV	Hole void visible
BVV	Broken void visible	ID	Infiltration dripper
CC	Crack circumferential	IR	Infiltration runner
CL	Crack longitudinal	JSM	Joint separated medium
CM	Crack multiple	OBR	Obstacle rocks
CS	Crack spiral	RBB	Roots ball barrel
D	Deformed	RBJ	Roots ball joint
DAE	Deposits attached encrustation	RFB	Roots fine barrel
DAGS	Deposits attached grease	RFJ	Roots fine joint
DNF	Deposits ingressed fine	SAM	Surface aggregate missing
DNGV	Deposits ingressed gravel	SAP	Surface aggregate projecting
FC	Fracture circumferential	SAV	Surface aggregate visible
FL	Fracture longitudinal	VC	Vermin cockroach
FM	Fracture multiple	VR	Vermin rat
H	Hole	XP	Collapse pipe sewer

FIGURE 3.5 Selected PACP defect codes, including commonly used structural and O&M observations. When defects are coded to a standard, more complex analyses, such as infiltration estimates, are possible. Public Domain: https://nepis.epa.gov/Exe/ZyPDF.cgi/P1008C45.PDF?Dockey=P1008C45.PDF

think rehabilitation is appropriate – sometimes threatening to withhold payment until a "correction" is made. While it is true that errors can occur in PACP coding (more on this later), if the coder believes the coding is accurate, codes should not be changed to make it seem like repair could be put off. Incorrectly downplaying the severity of defects can put lives at risk, and inspection contractors should not succumb to this pressure. If a customer wants to override codes and take on additional liability, let them make that decision. Don't make yourself a scapegoat.

Continuous defects are addressed algebraically. The length of a continuous defect in feet divided by 5 (or 1.5 if meters are used) will output an equivalent number of point defects. Decimals are rounded to the nearest whole number. For example, a continuous defect extending for 42.3 feet would translate to 42.3/5 = 8.46 or 8 equivalent point defects. While the methodology behind this system is empirical, it provides a standard for evaluating continuous defects and ensures that a quantitative assessment does not become a judgment call. Further, it ensures that calculations are standardized from assessment to assessment.

PACP also provides several mechanisms to score an overall line based on the number, severity, and type of defects found within the line. These mechanisms include quick scores, segment grade scores, overall pipe ratings, and pipe rating indices. The quick score is the easiest to calculate – it is a four-digit number, analogous to a binary-coded decimal, in that each digit has a special representation, not tied to traditional place value. The first digit represents the highest severity grade, and the second digit represents the number of defects with that severity grade. The third digit represents the next highest severity grade, and accordingly, the fourth digit represents the number of defects with that severity grade. For example, a pipeline with a quick score of 4321 would represent a pipeline with three defects receiving a grade of 4, and one defect receiving a grade of 2. There are no defects with grades of 5 or 3. A pipe segment with no defects will have a score of 0000.

When more than nine defects are observed, letters are used in the second and fourth digits. Each letter represents a numerical range of 5 defects. A represents 10–14 defects, B represents 15–19 defects, and so on. Thus, a pipeline with a quick score of 5A4F would have 10–14 defects with a grade of 5 and 30–34 defects with a grade of 4 – it is likely in need of some urgent corrective action! Remember, that while these quick ratings may resemble hexadecimal numbers, they are unique to PACP and should not be conflated. Many software packages will automatically generate this quick score upon completion of an inspection.

More detailed scores are possible. Segment scores provide another way of evaluating the O&M and structural condition of pipe segments. These segment grades (SGs) exist for each score level and can be calculated by multiplying the number of defects by the score of each defect. Since a segment grade exists for each score, this simply provides a weighted count of defects and is not widely used. For example, consider a pipe with the following defects:

Condition Grade	Structural Defect Count	O&M Defect Count
1	3	4
2	5	5
3	2	1
4	2	5
5	1	0

The structural grade would be as follows: $SG_1=3\times1=3$, $SG_2=5\times2=10$, $SG_3=2\times3=6$, $SG_4=2\times4=8$, and $SG_5=1\times5=5$. O&M structural grades would be calculated similarly. $SG_1=4\times1=4$, $SG_2=5\times2=10$, $SG_3=1\times3=3$, $SG_4=5\times4=20$, and $SG_5=0\times5=0$. Again, since structural grades exist for each score, weighting obfuscates the useful information represented within: a defect count. While many software packages will generate these statistics as part of their PACP compliance, it is more useful to simply use defect counts to analyze and plan for rehabilitation.

Structural grades do have utility when they are fed into an overall pipe rating (OR). The OR consists of the sum of all of the structural grades, and is reported separately for structural and O&M evaluations. In this case, weighting adds needed value in amplifying the severity of defects. An argument could be made that the linear scale adopted by PACP is not appropriate (should the jump from 3 to 4 be the same weight as a jump from 4 to 5?), but as with much in civil engineering, it provides an empirical set of standard calculations that facilitate comparison between datasets. In the example above, the Structural OR would be $3+10+6+8+5=32$, and the O&M OR would be $4+10+3+20+0=37$. Higher ORs indicate lines more in need of corrective action, after length is taken into account.

Finally, the pipe rating index (RI) provides a summary statistic that describes the severity of an average defect in a segment. A pipe could have a high OR number by having a large quantity of mild defects or a smaller quantity of severe defects. The RI expresses this by dividing the OR by the total number of defects. Again, using the example above, the Structural RI would be 32/13=2.5, and the O&M RI would be 37/15=2.5. Note that the OR and RI scores can be useful in the sense that they provide a snapshot of the overall condition of the pipeline, but also mask information on the most severe defects. If 90% of a pipeline is fine, but 5% has a critical failure that could endanger the health and well-being of those in the vicinity, it is likely in need of immediate attention. The Quick Rating provides that information in an easy-to-digest format, and thus is used more widely than the other scores.

While these grades provide an easy way to calculate some useful summary statistics, they should not replace the assessment of engineering professionals. More specifically, while PACP condition scores are useful tools in making repair and rehabilitation decisions, they should not be the only factors – all too often, municipalities will come up with a plan to address all the defects rated 4 or 5 without making a comprehensive plan. Some severe defects can be stable for months or years depending on service conditions, while moderate defects can accelerate quickly if conditions are right. Does it make sense to perform multiple point repairs or plan to reline a pipe segment? What is the schedule for road paving in the area? Could disruption to traffic be minimized by coordinating efforts? Best practices would suggest that filtering by defect severity can be a useful tool in identifying lines in need of additional rehabilitation, but should not be the sole tool used in making specific rehabilitation decisions.

In more recent versions of PACP, some risk analysis tools are also provided. Specifically, PACP provides metrics for calculating the likelihood of failure (LoF) and consequence of failure (CoF), the product of which is one method of quantitatively expressing risk. PACP's criteria for both are expressed on a scale of 1–6. LoF is calculated by dividing the first two digits of the quick score by 10. For example, in a pipe with a quick score of 3522, the LoF would be 35/10=3.5. If the second digit is a letter, it is replaced with a 0 for the division and 1.0 is added to the final score. For example, 4A15 would have an LoF of 40/10+1.0=5.0. If no defects are present, the line receives an LoF of 1.0. The maximum LoF is 6.0.

Consequence of failure can be calculated in a variety of different ways depending on the cost a failed line would impose on the community, environment, and bottom line. PACP provides several suggestions for assigning lines a score of 1–6 depending on their size, proximity to sensitive natural areas, depth underground, class of surrounding roads, and ease of access. Much of this information must be obtained by tying inspection data to GIS data and information from the rest of the system. It cannot be determined from inspection data alone. Responsible risk management programs would assign a risk to each pipeline by multiplying the LoF and the CoF, and devising a strategy for each.

Many EPA consent decrees and other federal regulations require municipalities to develop their own asset and risk management programs. Compliance with best practices for capacity management operation and maintenance (CMOM) programs is key to demonstrating progress to the EPA, and the Governmental Accounting Standards Board Statement 34 (GASB 34) requires municipalities to include infrastructure assets in their financial reporting of all assets. Thus, it is essential that even smaller operators of wastewater collection systems develop some kind of quantitative risk management strategy.

While it is easy to poke holes at the rudimentary approach PACP takes to this problem, it is a suitable introduction to the topic. For a municipality with limited resources, the PACP approach is a way to begin a data-driven approach to asset management, using a simplified, easy to understand calculation system, the details of which come with a training course. While large municipalities and those with the budget to hire engineering partners and consultants to run their systems may have much more sophisticated models, the PACP approach can be very valuable to smaller entities looking for a cost-effective point of entry into the space.

SOFTWARE PACKAGES

PACP provides a standardized way to inspect pipes, and while it could be implemented on paper forms, the vast majority of users inspect pipelines using some kind of integrated software package. These packages include different modules for data collection and office analysis, and are available from a variety of different vendors. Leading vendors include WinCan, PipeLogix, RedZone Robotics (ICOM), Cues (GraniteNet), and IT Pipes. Each of these software packages went through a certification process developed by NASSCO in order to ensure that they conform to the PACP 7.0 standard and output standardized data files.

The production of standardized data files is extremely significant in vendor interoperability. Prior to PACP version 7, in many markets WinCan was a de facto standard – it was specified into contracts, was widely installed by inspection contractors, and provided a well-understood interface for analyzing data from multiple projects. Unfortunately, as other software developers began to establish feature parity with WinCan, and even went beyond the offered functionality, effective data exchange was difficult to achieve with WinCan's proprietary format.

Now that most software packages are able to import and export data in the PACP standard format, vendor lock-in becomes much more difficult and competition continues to improve the quality of data coding platforms. Further, a variety of products are available to suit a number of applications and budgets: some inexpensive software packages are designed for standalone use on CCTV trucks. Other more expensive products are structured like software-as-a-service (SaaS) platforms with annual licensing fees, web interfaces, GIS integration, and the ability to work with large datasets.

While this competition benefits the industry, some caution is essential. The PACP database standards were developed by sewer professionals – not programmers. As a result, the initial versions of PACP 7.0 included enough ambiguity that while all packages handled test data in the same way, they produced a variety of outputs with real-world datasets, meaning that older versions of WinCan data would not always be imported into ICOM without issue (for example). As NASSCO and its software vendors continue to improve standardization and data interoperability, these issues will become less and less significant.

The specific format for data exchange in PACP 7.0 is a database in Microsoft Access (.mdb) format, with a variety of predefined tables. Images of defects are stored as JPEGs in a particular folder structure and the associated video file can be in a variety of formats. This data structure has tables that are named in intelligible ways, and NASSCO makes various data dictionaries available to members upon request. It is worth familiarizing oneself with this format, as many minor hiccups can be resolved with a few simple database edits. Since the data is stored in a Microsoft Access format, this can be accomplished through a graphical user interface and does not require the knowledge of structured query languages like SQL.

This is significant as technology continues to improve, and various inspection software packages will invariably add features faster than the NASSCO standard can catch up. New forms of video encoding can sometimes break the database references – meaning when users click on a defect, the software package will fail to jump to the correct location in the video to observe that defect. Learning to edit these databases makes it possible to correct those issues and take advantage of compression technologies that improve video quality and simultaneously reduce file size.

Further, many municipalities will use larger-scale Enterprise Asset Management (EAM) software packages – tools like CityWorks, Hansen, or IBM's Maximo – or geographic information systems that would benefit from integration with inspection data. Some knowledge of databases, table structure, and exchange functionality can go a long way toward resolving compatibility issues from multiple data sources, and produce processes that can save utilities millions of dollars. For example, by integrating wastewater inspection data with an EAM and GIS, work crews could be scheduled to repair several observed defects in a sewer line, locate those defects within a meter of accuracy, and record the completion time and cost of the repairs in one unified location. The inspection and asset records could also be updated to note that the defect was addressed. More details about data formats and integration will be discussed in subsequent chapters.

Thus, many municipalities will specify that data be collected in any NASSCO PACP compliant tool – doing so allows inspection contractors to use a variety of software to compete for work and drive down prices. However, many tools provide time-saving administrative interfaces, and there is some value in specifying that inspections be collected in the same program that the municipality or engineering firm will use to review the data. Remember, a contractor wanting to be as competitive as possible will need to employ software from multiple vendors municipal customers should specify whatever software they are comfortable with and put the onus on contractors to handle the conversion. Settling for a compromise in data formats could lead to pesky interoperability issues that take months or years to resolve.

PRACTICAL ISSUES WITH PACP INSPECTIONS: DATA QUALITY

In the United States, PACP has become a standard in the pipeline inspection industry, such that public works directors, wastewater engineers, and inspection contractors are all familiar with the standard at some level. Yet, many projects are delayed and deliverables are rejected because of data quality issues. On some level, this will always be an issue: any married person can attest that two reasonable people can look at the same image and see two very different things. As a visual inspection technology, PACP is no different, though its well-defined defect descriptions and standardized code set the limits of true debate to a few edge cases. Problems start to arise when poor quality video is submitted for coding, or when operators make significant errors.

Video quality can be a significant problem in accurately coding pipelines. If a video is too dark or oversaturated, it can be difficult to accurately see and classify roots, cracks, and intrusions. As reinspection can be a difficult and expensive process for contractors, many (all?) will advocate strongly for the acceptance of flawed inspection data as a necessary cost of performing pipeline inspection in the real world. This can be true to an extent: insisting on inspecting sewage lines in below freezing weather can result in fog that must be removed with blowers (Figure 3.7). Trying to inspect a pipe without cleaning can result in debris obscuring part of the lens.

That said, many video quality issues can be easily avoided. Requiring operators to perform a lighting and video adjustment at the start of every inspection can go a long way to clearing up video quality issues, as many simply forget to adjust the camera when they begin a new inspection – even if they are working in pipe of a different size or material. Including in the specification that each inspection must include a 10–15 second video, visible video adjustment period prior to beginning forces operators to think about video quality, make some adjustments, and consider the overall view before pushing ahead with an inspection (Figure 3.6).

This cannot always be corrected in the post-processing stage. While software packages like Apple's Final Cut Pro and Adobe Premiere can adjust the brightness and contrast of a variety of video files, making these changes after inspection to drive acceptance of inspection footage misses the point. If the video quality was substandard during inspection, the coder used that video to recognize features, stop and examine them, and classify each appropriately. Just because an inspection video can be made to look bright in the post-processing stage doesn't change the fact that the inspector running the robot in the dark would have been unable to see small cracks or defects in the pipe. Spotting those defects in pre-recorded video without the ability to PTZ on demand becomes guesswork at best.

Thus, many inspection contracts get hung up prior to closure – when customers reject data collected with poor lighting. Contractors claim that they should be compensated since they did the hard work of inspecting difficult-to-access lines. Municipalities push back, saying that since they can't feel confident that all defects were caught a reinspection is in order. Inspection contractors should train crews to prioritize data collection – which goes well beyond the act of getting the robot from one manhole to the next. Files should be checked during and immediately after an inspection for quality and completeness, and any issues should be addressed before leaving the job site. It is usually far cheaper to adjust lighting and rerun the robot before leaving the job site than it is to return weeks or months

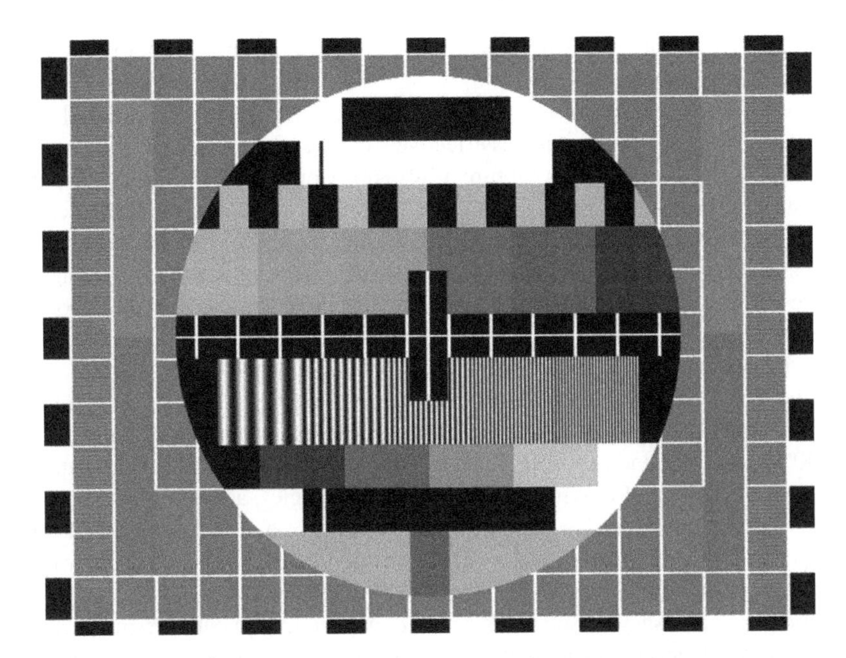

FIGURE 3.6 PM5544 is an example of a test pattern, typically used in European analog television systems. Printing out a few copies of this and laminating them is an easy way to produce test cards that will verify camera function at the start of each inspection. Starting with a quick scan of the card will only take a few seconds and can prevent hours of rework later. Public Domain: https://commons.wikimedia.org/wiki/File:PM5544_with_non-PAL_signals.png

FIGURE 3.7 While some may claim that fog in a wastewater pipe is inevitable on a cold day, blowers such as the one shown above and adequate ventilation can prevent it from obscuring CCTV inspection data. It does require planning, and sometimes more extensive traffic control. Public Domain: https://www.osha.gov/SLTC/etools/shipyard/shipbuilding/confinedspace/ventilation.html

later and start the process afresh, with traffic control, permits, and all else that is required with sitework.

Fisheye camera inspections have an added advantage over traditional video inspections, as many lighting issues effectively can be corrected in the post-processing stage. Since fisheye robots capture a full 360° view of the interior of the pipe, it is possible to virtually drive the robot down the pipe again and reinspect on demand. Thus if a video is too dark, it can be lightened (to a point), and an inspector can reassess the condition as though he were in the field. Many contractors leverage this feature to improve coding quality. Inspection crews will simply drive the robot from manhole to manhole, ensuring that the lenses don't get occluded. Dedicated office staff will then complete the laborious process of coding each defect.

This division of labor tends to improve inspection efficiency and generate higher quality coding. Field workers can inspect much larger quantities of pipes – think 10,000′ in a day instead of 1,000′ in a day. Office workers get better at coding, since they have the luxury of looking at a defect from all angles, adjusting the brightness, and getting a comprehensive view from the comfort of a climate controlled office. Remember, live coding often takes place in the middle of the night, along busy roads, during all types of inclement weather.

Unfortunately, fisheye inspections have a downside from a data perspective: there is no universally accepted standard for these virtual PTZ videos. Software packages can export traditional movies, but giving the user the ability to pan, tilt, and zoom around the pipe on demand often requires the installation of a special viewer or accessing a custom website. This functionality is also unsupported by EAM software packages or other upstream integration targets. Further, if fisheye inspections do not have sufficient resolution to capture the pipe interior in detail, no amount of ex post facto correction can make up for this. It is impossible to generate pixels from nothing, and blocky, blurry images may not provide an acceptable level of detail for inspection reviewers.

Regardless of whether or not inspections are collected with traditional PTZ cameras or fisheye systems, all defect codes should be spot checked or reviewed by inspectors or supervisors. Coding sewer pipelines is boring, and it is normal for individuals to miss the occasional code. However, it is also easy for one person to misunderstand some aspect of his training and miscode defects in tens of thousands of feet of pipeline. In traditional PTZ systems, that means potentially tens of thousands of feet of rework.

A good rule of thumb is to check 10% of all inspections performed by each coder on a weekly or monthly basis. If all is well, give them some positive feedback or maybe a data quality bonus and move on. If corrections need to be made, address them with the coder in question – show what is wrong with an incorrect code, and what it should be changed to, along with an explanation of the rationale for why these changes should be made. Inspectors will also find value in going out to a job site, and watching coding as it occurs to fix issues in real time. Quality control issues with PACP coding account for a significant amount of rejected deliverables and rework, and investing some time in improving code accuracy will pay dividends during the contract close-out process.

NASSCO should consider making some changes to their PACP certification process in order to improve data quality, as well. At the end of the MSCC certification course, candidates must watch an actual video of a sewer inspection and code all of the defects they see. Completing this task accurately is key to earning a passing score on the exam. By contrast, PACP asks its candidates to answer a variety of multiple choice questions about coding in an open book exam format – a task that seems reasonable until one realizes that the actual images and codes in question are presented in the course textbook. Students can simply look up each photo and pass the exam without demonstrating an ability to code observed defects. Requiring students to accurately code a video they haven't seen before would be a much more useful demonstration of ability.

Neither MSCC nor PACP have students complete these tasks electronically – so much attention on the exam is given to opening and closing continuous defects using the proper nomenclature, even though most software packages address this automatically for end users. Some improvements to modernize the certification process could dramatically improve the performance of defect coders in the field. Ideally, an exam would involve students adjusting lighting, looking at defects properly, coding them accurately, and repeating this process for the length of an inspection.

Finally, this may sound obvious, but is worth noting: any coding system is only capable of recording defects that are visible to a camera or the naked eye. This means that defects that are obscured by flow or masked by other defects cannot accurately be represented. A visual inspection may show unusually high flow levels, but it is not the place of a CCTV inspection to speculate whether that is due to debris below the flow line or unusual conditions elsewhere in the system. Further, a defect may be visible, but its severity may not be quantifiable using any combination of codes in a CCTV inspection.

For example, hydrogen sulfide gas is capable of dissolving many types of reinforced concrete pipelines. It eats away through the outer surface of concrete, producing soft pockets of gypsum which crumble away, leaving additional interior surfaces to corrode even more. This corrosion is almost always visible to the naked eye in the form of visible or protruding aggregate particles or other indicators of the interior of concrete structures. However, the presence of visible aggregate cannot accurately indicate how much corrosion has occurred. In a 4″ thick pipe, does visible aggregate mean that the surface has only begun to corrode or that massive wall loss has occurred and failure is imminent? Remember, rebar might not be visible if it has been completely eaten away. Sometimes clues can be obtained by reducing flow levels, and looking at the wall loss above and below the flow (hydrogen sulfide corrosion can only occur in headspace), but sometimes additional sensor technologies will be required to conclusively answer these questions with a degree of quantitative certainty.

As of now, PACP does not include provisions for integrating data from other sensors such as sonar, laser, or hydrogen sulfide gas. However, operators can always insert general observations (MGO codes) with readings of these sensors at various locations or at features of interest. For example, if laser data is available, it may be useful to log a general observation of diameter measurements at the same location at which rebar is visible.

4 Quantifying Condition above and below the Flow

An Overview of Sonar and Laser Technologies

While visual inspection technology is one of the predominant tools used in condition assessment today, it has two major shortcomings: first, despite the inclusion of NASSCO's rating system, it is a qualitative system. While roots may be classified as small, medium, or large, and receive a corresponding numerical grade, CCTV inspections cannot effectively measure the diameter of each root. Similarly, the length of cracks or the amount of wall loss cannot be computed from a visual inspection – all such an inspection can do is note the presence or absence of defects. Second, visual inspection technology is limited to observations above the flow. Most flows in buried pipelines have significant amounts of turbidity and are generally opaque. Therefore, CCTV cameras cannot observe the presence of debris unless it is protruding from the flow line.

Most coding systems take this one step further and explicitly instruct inspectors to not code things that they cannot see. If a crawler rises during an inspection as it is traveling through a pipeline, it is safe to say that it has passed over some type of debris. Yet, unless those debris are visible, they will not be assigned a defect code, aside from perhaps a general observation. To do otherwise would be to engage in guesswork. What do those debris consist of – settled solids, missing pieces of aggregate, large grease deposits? Any code would be making some kind of assumption, and could over- or underestimate the severity of the problem.

To solve this problem, many sewer inspections use additional sensors that go beyond the capability of CCTV. To "see" below the flow line and quantify debris, high-frequency acoustic signals are used. Sonar transducers send and receive these kHz-MHz frequencies and produce a picture of the debris profile below the flow. Above the flow, different types of laser technologies are used to measure the inside diameter of the pipe at various locations and estimate the amount of wall loss, corrosion, or buildup that has taken place. Current technologies involve the use of visible light ring lasers or infrared lasers used in time-of-flight (lidar) systems.

PROFILING SONAR

In 1490, Leonardo da Vinci inserted a pipe into a body of water and listened carefully. He determined that he was able to detect the presence of passing objects by listening to changes in the sounds that were produced. Unfortunately, these qualitative observations were not able to become part of a quantitative technique until the late 1800s, when the underwater speed of sound was determined, and underwater acoustics became useful to the maritime industry. In 1906, Lewis Nixon invented a detection device that listened for changes in underwater acoustics in order to detect icebergs and help ships chart clear courses through North Atlantic shipping lanes. After the Titanic sunk in 1912, inventors scrambled to come up with ways to locate the wreck using more active underwater echolocation systems, in which sound waves were transmitted through the water and reflected back upon contact with an object. The elapsed time between transmission and receipt correlated to the distance the sound traveled (Figure 4.1).

World War I triggered more massive research and development efforts, when submarine warfare devastated surface fleets. Developing systems that could locate and trace submarines was key to achieving naval superiority. In 1915, Paul Langévin and Constantin Chilowski created a device that sent and received audio signals in order

FIGURE 4.1 Some of the earliest studies of hydrophones were used to determine the speed of light underwater. These same principles form the core of modern sonar. This image depicts the first recorded experiment in 1826. Public Domain: https://dosits.org/people-and-sound/history-of-underwater-acoustics/the-first-studies-of-underwater-acoustics-the-1800s/

to detect submarines. Their device could be mounted on surface ships, and quickly displayed the location of enemy vessels far below the surface.

While all of these examples could be considered precursors or relatives of sonar, the term itself was created and employed by the United States in World War II. The British referred to early sonar-type devices as ASDICs, stemming from the Anti-Submarine Division responsible for the R&D program in question. The Americans needed a term that would quickly indicate the function of these acoustic devices to seamen coming to terms with a wide array of jargon and abbreviations. Sonar – an acronym/portmanteau stemming from sound, navigation, and ranging – was chosen because of its similar functionality to radar. Sonar detects encroaching objects underwater just as radar detects them in the air (Figure 4.2).

Sonar systems can be broken down into two broad categories: passive and active systems. Passive sonar systems work similarly to da Vinci's tube – devices such as hydrophones are placed in a fluid and listen for changes in order to detect the presence of objects (Figure 4.4). Matching acoustic signatures can even be employed to identify what the objects are. Active sonar systems involve both the transmission and receipt of sound, and are the predominant form of sonar attached to moving vehicles, as they are able to overcome the ambient noise present in operation.

Sonar devices used in underground pipelines are active sonar devices, and their functionality is remarkably unchanged from the principles established in the first half of the 20th century. The active sonar process begins when a tone or pulse is generated, amplified, and transmitted through a transducer. While submarine movies often employ an audible "ping" to indicate this process, most pipe profiling sonar units use ultrasound – frequencies in the thousands or millions of hertz – that would

FIGURE 4.2 It should come as no surprise that the United States Navy has done much to foster the development of sonar technology. This image is of sonar technicians at work aboard a nuclear submarine. Public Domain: https://commons.wikimedia.org/wiki/File:Sonar.jpg

be inaudible to the human ear. Higher frequency sounds can be focused into a narrower beam, while lower frequency sounds travel longer distances through a fluid (without distortion). In buried pipelines, sound waves need to travel a few hundred feet (at most) and accurately measure the shape of debris. Ultrasonic waves are able to be formed into beams that can detect half an inch of debris a few feet away from a crawler, and are the most appropriate frequency range for the application. Typical frequencies range from 600 kHz to 2 MHz.

The sonar transducer is attached to a stepper motor that is responsible for the "sweep" of the sonar head. The stepper motor rotates around a central axis 360° so that the audio beam can be sent around the entire circumference of the pipe. There is a tradeoff between scan speed and resolution – faster rotation "smudges" the beam, so that detailed sonar scans are a relatively slow process. One full sweep may take just under a full second to complete – vendors don't like to advertise this fact, as 1 Hz does not come across as a compelling specification! Each sweep corresponds to a complete cross-sectional snapshot of the pipeline's interior.

Finally, the reflected audio signal is received by a hydrophone tuned to process the frequency that has been transmitted in order to produce a low noise display of debris (Figure 4.3). More expensive systems can employ an array of hydrophones, but simple, inexpensive units used in pipelines can use a single sensor. An onboard processor keeps track of the angle of transmission so that received signals can be stitched together into an accurate cross-sectional representation. Since sound travels over 1400 m/s in water and sweeps are relatively slow, there is little chance of signal overlap over the distances being observed in wastewater lines.

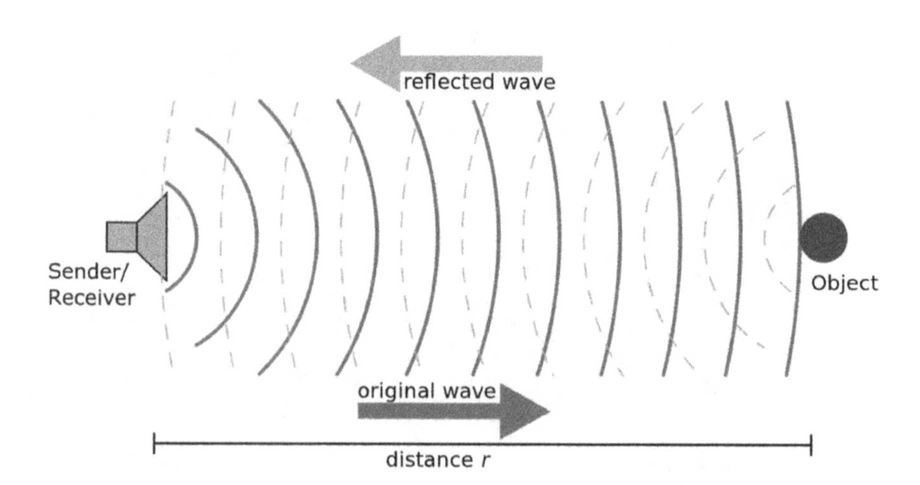

FIGURE 4.3 The fundamental principle of sonar involves sending out a pulse and measuring the amount of time it takes to hear an echo. This pulse does not need to be an audible ping – frequencies outside the range of human hearing are often selected for signal quality and detection thresholds. Image provided by Georg Wiora. Creative Commons: https://commons.wikimedia.org/wiki/File:Sonar_Principle-sv.svg

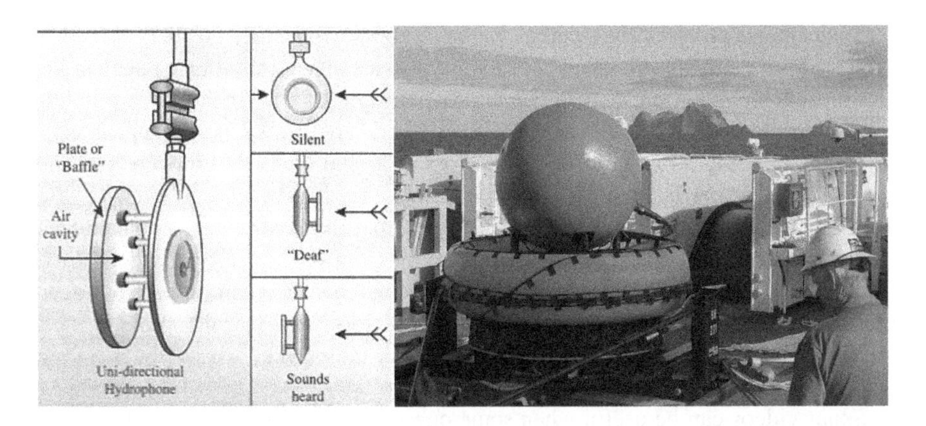

FIGURE 4.4 Hydrophones are sensors that translate vibration into an electrical signal, though the process used can vary based on size and application. Sometimes physical chambers of air or fluid are used, while piezoelectric crystals can provide a much more precise measurement. A schematic representation is shown at left, and an example of a large hydrophone for oceanic deployment. Schematic drawing provided by W. Van der Kloot and photo of hydrophone provided by the Comprehensive Nuclear Test-Ban Treaty Organization. Creative Commons: https://commons.wikimedia.org/wiki/File:Hydrophone_drawing.jpg, https://commons.wikimedia.org/wiki/File:A_hydrophone_node_(13466196874).jpg

A variety of vendors produce and sell profiling sonar units (sometimes referred to colloquially as transducers, even though they include both transducers and receivers) that communicate through standard serial or Ethernet interfaces. Examples include Imagenex, Inc. and Marine Electronics. Each unit features slightly different ultrasonic frequencies, scan rates, beam diameters, and physical housings, so some level of feature analysis may be required when choosing a sensor for a specific application. Vendors also provide turnkey sonar inspection systems – with a sonar head mounted on a robotic crawler or floating platform. RedZone Robotics HDSub and Cues Inc.'s Sonar Profiler are standalone units ready for deployment – minimal systems integration is required. From the inspection specification perspective, a profiling sonar unit capable of operating in the pipe diameters in the system is usually sufficient. Be careful – vendors insisting on the insertion of language specifying a 600 kHz ultrasonic beam (for example) are typically trying to exclude competitors. It would be difficult to find a pipe profiling sonar unit in use today that has a fundamental technical incompatibility with the goals of an inspection program.

Unfortunately, there is no standard format for sonar inspection data. Most sensors stream data over serial or Ethernet, and it is up to the device manufacturer or system integrator to decide on an appropriate format for storing this data. The most efficient way to do so is in a machine-readable binary format – which is not particularly useful to human users. A more useful format is the time-indexed comma separated value file which can be easily imported into a spreadsheet. Beware, however – it is easy to generate very large file sizes using this format, so it may not be appropriate for representing entire inspections.

Some manufacturers will output "sonar movies" in which the data stream is played back in a movie file. This looks similar to a sonar installation as depicted by Hollywood: a circle represents the pipeline's interior, and a beam sweeps around the interior similar to the hands on a clock. When debris are located, points appear as the beam passes, generating a profile of the cross-sectional interior of the pipe one second at a time. Many feel a sense of comfort when viewing these sonar videos – after all, they are the closest thing to tangible proof that the contractor completed their inspection and analyzed the sonar data correctly. Comfort aside, sonar videos are of limited analytical utility – the video format means that data can only be extracted through computer vision techniques, and contractors can typically provide much more useful summary statistics, such as graphs or tables of debris height as a function of chainage.

Sonar videos can be useful when some question about the substance of debris is under consideration. For example, a sharp profile can be indicative of rigid debris, while a fuzzy profile can show softer debris – such as settled sediment. Thus, if an inspection must be abandoned in a line in the vicinity of a cement plant, the sonar profile can indicate whether the obstruction consists of sediment that could be washed away through jetting, or solid concrete that will require much more drastic measures. Sometimes, other questions can be easily answered by looking at sonar videos – if displaced bricks are observed in the crown of a pipe, sonar videos will often show their silhouettes a few feet downstream below the flow. Contractors who complain about an inspector forcing them to redo an inspection unnecessarily may have data on their side if the tread prints from the previous inspection are visible in the debris profile.

Sonar videos can be colorized in post processing, and many vendors offer this as a value-added feature. In these colorized videos, the pipe wall is overlaid as a circle, and debris protruding from that circle are assigned a different color based on height. For example, debris that are 3″ above the invert may be represented as red, whereas debris that are less than an 1″ in height may be colored green. This coloration can be a useful decision support tool, but know that the thresholds and centering are typically set manually by the inspection vendor. Just because a pipe has no "red" debris does not mean that there aren't areas of concern – it could just mean the inspection contractor has chosen to set the threshold for what should be colored red at a different location than the project engineer. Different inspections may require different thresholds – debris may be much less significant in a cleaning analysis than in a pre-lining inspection. If a data reviewer does not like the coloration applied to sonar video data, they should feel comfortable requesting that the contractor make changes.

Within the context of water and wastewater pipe inspections, profiling sonar data can be used for two key purposes: debris quantification and ovality measurements. Debris quantification is a relatively straightforward process. As an inspection occurs, the height of debris can be measured by the sonar sensor. If these debris heights are integrated over the inspection distance, the volume of debris in the line can be calculated – a much more useful quantity. Since the EPA requires asset owners to maximize the available capacity, the effectiveness of a cleaning program can be validated by small debris volumes. Further, since cleaning contracts are often priced based on

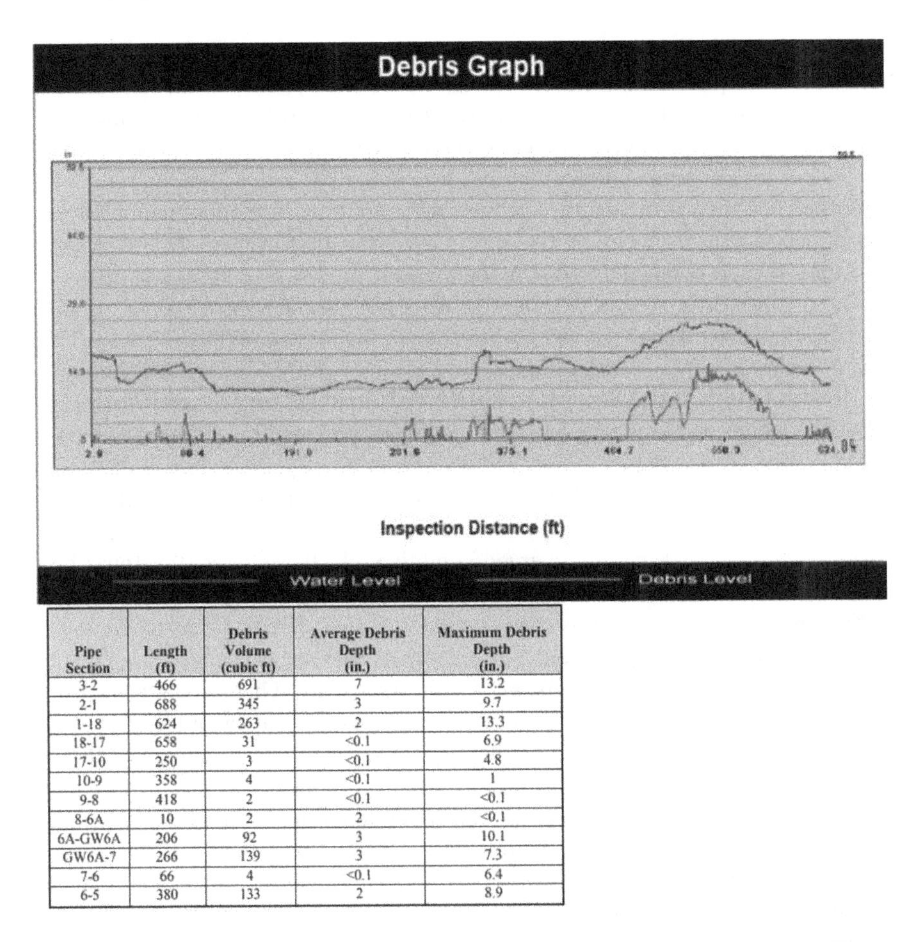

Pipe Section	Length (ft)	Debris Volume (cubic ft)	Average Debris Depth (in.)	Maximum Debris Depth (in.)
3-2	466	691	7	13.2
2-1	688	345	3	9.7
1-18	624	263	2	13.3
18-17	658	31	<0.1	6.9
17-10	250	3	<0.1	4.8
10-9	358	4	<0.1	1
9-8	418	2	<0.1	<0.1
8-6A	10	2	2	<0.1
6A-GW6A	206	92	3	10.1
GW6A-7	266	139	3	7.3
7-6	66	4	<0.1	6.4
6-5	380	133	2	8.9

FIGURE 4.5 An example of the presentation of sonar data. The graph indicates variations in debris and water height during the inspection. Remember, that the sonar head must always be submerged in order to generate data. Most vendors will provide a summary table in addition to the graphical summary. The debris graph is for a single pipe segment, while the summary table contains statistics from multiple inspections. These examples are taken from the EPA's Report on Field Demonstration of Condition Assessment Technologies from 2011. Public Domain: https://nepis.epa.gov/Exe/ZyPDF.cgi/P100C8E0.PDF?Dockey=P100C8E0.PDF

the volume of debris removed, sonar measurements of debris volumes can provide accurate estimation and a check on contractor provided quantities (Figure 4.5).

However, these debris estimates should be thought of as rough quantities, as calculus has not made its way to many sonar analysis software programs. Debris are often calculated manually, using rough geometric approximations. For example, a debris profile may be represented on a grid. Counting the squares under that profile gives a rough approximation of the cross-sectional area occupied by debris. If this process is repeated for every sonar scan and multiplied by the distance between scans, an approximation of debris volume can be obtained.

This, coupled with the low scanning rate of commercially available sonar heads, leads to debris quantification that is best thought of as approximate. Just because a contractor's report estimates that there are 34,127.3 ft^3 of debris, does not mean that any party involved in the inspection has confidence in the last decimal place. Instead, planning for a cleaning contract that will necessitate the removal of approximately 30,000 ft^3 of debris is an appropriate conclusion to draw. Too often, data reviewers and project engineers focus on the least significant digit in their data acceptance decisions – this is completely missing the mark.

These debris estimates provide the real value behind sonar inspections, and some thought should be given to the visualization and utility of data. While the overall volume of debris in a line is a useful quantity, a visual map showing the distribution of debris within the pipe can be helpful. If debris accumulates immediately after a specific tap, that can be a sign of a damaged lateral and potential inflow source. Further, spatially representing debris in pre- and post-cleaning inspections can make a compelling and obvious case for whether or not the cleaning operation was sufficient.

In cast iron pipes, sonar can also give a sense of tuberculation or ovality, especially when employed in fully charged conditions. In these cases, the sonar head will be able to observe the complete interior of a pipeline – from crown to invert, and all areas in between. In this case, the amount of scale buildup or tuberculation can be measured, but keep in mind that sonar does not have the accuracy needed for precision measurement, such as that needed for liner sizing. On most units, measurements close to the sonar head (< 12 ft away) can have an accuracy of less than 0.25″. At larger distances, that accuracy can fall to 0.5″ or worse. This combined with the slow scan speed makes sonar a useful tool for drawing broad conclusions. A sonar inspection of a fully charged pipe can inform an asset owner that there is roughly 1.5″ of scale buildup throughout the line, but would not be able to accurately say at what point the buildup precisely reaches a level of 2.7″ exactly, unless a specialized inspection were undertaken.

Similarly, in flexible pipes, sonar can be an excellent tool for observing ovality and deflection. However, care must be applied to ensure that the data collection device is pointed directly down the axis of the pipe, as slight misalignment can cause scans to appear wider than they are tall – producing false indications of ovality. Calibrating the device before beginning an inspection can help alleviate this. If several stationary scans are taken before the device begins to move, a baseline of the vertical and horizontal diameter of the pipe can be obtained. True ovality would see a reduction in the vertical diameter and an increase in the horizontal diameter, while a misaligned sonar head would show only an increase in the horizontal diameter while the vertical diameter remains constant.

Inspection of fully charged pipes can involve either a crawler moving along the invert or a neutrally buoyant submersible that floats throughout the inspection. Each device has specific strengths and weaknesses. The crawler tends to move much more slowly, but can produce higher-resolution sonar data, as a heavier robot is less likely to be buffeted by flow (Figure 4.6). It is also more difficult for a crawler to flip unexpectedly. Submersible devices can inspect lines in much shorter periods; however, they can flip multiple times, and knowing the orientation of the device can be

FIGURE 4.6 Sonar heads are often mounted underneath inspection platforms, where they will reliably remain submerged. This shows an example of a profiling sonar unit manufactured by Marine Electronics, Inc. Use caution when transporting – this low mounting point means it is easy for sonar heads to inadvertently take the brunt of impacts, and they are often easily damaged. Consider protecting the device with a physical cage. Public Domain: https://nepis.epa.gov/Exe/ZyPDF.cgi/P100C8E0.PDF?Dockey=P100C8E0.PDF

difficult to determine from sonar data alone. Air pockets in the crown can look very similar to debris in the invert.

The solution is to insist that vendors include data from the device's inertial measurement unit, or IMU. The IMU is essentially a set of gyroscopes that reports information on the device's orientation and acceleration. IMU data has several significant issues – notably drift – that should be accounted for if it is to be relied upon for precise analyses. However, to check sonar data, one only needs to confirm the direction of gravity – e.g. which way is down – in order to confirm that debris are truly debris. This is useful data to obtain for a crawler inspection as well, as well-meaning data processing technicians can sometimes flip the crown and the invert – especially in clean pipes with an air pocket.

Several practical considerations should be noted for profiling sonar units. In order to accurately send ultrasonic signals into an aqueous medium, sonar heads must also be filled with fluid. When a sonar head runs out of fluid, no returns will be possible. Some can mistake the data from this malfunctioning device as a clean pipe. A good way to differentiate is to look for the telltale sweep, showing progress in time. Even clean pipes will show slight fluctuations in the sonar video as you move down the pipe. A dry head will not be able to produce or receive a suitable echo, and no change will result from scan to scan.

It is also possible for some flows to possess so much turbidity that sonar data collection is not possible. In this case, particles in the flow are large enough to reflect the ultrasonic wave back to the sonar head. While most flow is not transparent to the naked eye, flows meeting these criteria are generally more slurry than fluid and have a consistent amount of particulate matter floating in suspension. If this occurs, look for areas of inflow that could enable these conditions. Sometimes crawlers can "kick up" debris as they traverse through a pipeline. This transient behavior may be visible in the sonar data, but should not impact overall quantitative results.

One key point of consideration for sonar inspections is that commercially available heads are able to scan a wide range of pipe diameters, but must be switched into an appropriate mode in order to do so. Different vendors may use different terms to describe this feature, such as range scales. Switching appropriately becomes a consideration when a device is frequently swapped between pipes of a variety of diameters. If the range setting is off, it can be easy to tell. Sometimes the pipe walls will not be visible. Sometimes debris will appear out of nowhere. If something looks wrong, ask your inspection contractor to verify the settings.

Sonar videos can be displayed in a variety of formats – they can be burned to DVD, viewed in Windows Media Player, or even uploaded to YouTube (make sure you Like and Subscribe!). A particular method of displaying sonar information has become an informal standard. The Totally Integrated Sonar and Camera Inspection Technique (TISCIT) is touted as a unified method of integrating sonar and CCTV data. Don't let the acronym fool you – the methodology is simple. Sonar videos are synchronized with a CCTV inspection, and the sonar video is overlaid on the flow, allowing a viewer to simultaneously see above and below the flow while watching an inspection video. This unified platform makes it easy to do things like detect a displaced brick, or verify that an increase in flow levels are due to submerged debris at a particular location. However, inspectors will not assign defect codes to observations in sonar data – per PACP specifications.

Some inspection systems produce TISCIT videos by default – others provide separate data files. While TISCIT is interesting, it is a visualization aid at best, and requiring it in a specification serves to discourage innovation in camera systems. Anyone insisting on TISCIT over separate sonar and CCTV inspections is likely trying to give an edge to a preferred vendor. The same information can be easily gathered from any sonar data source, and TISCIT does not aid in debris quantification. Given the predominance of widescreen displays (that did not exist at the time of TISCIT's conception), a more useful view would be to show the sonar and CCTV videos side-by-side, allowing an unobstructed view of each. While this would not layer the sonar in an area of the screen corresponding to flow, it would enable vendors to upgrade camera and sonar systems independently, allowing technology to advance at a faster rate.

While pipe profiling sonar is the dominant technology in use today, other types of sonar may be appropriate for specialized applications, such as mapping a large chamber or other structure. Profiling sonar sends out its ultrasonic signal in the form of a conical beam. One return value or point in space is plotted for every sample taken, making it suitable for visualizing the profile or silhouette of debris. Imaging sonar units send out a beam that deviates by a fixed number of degrees above and below the center point. This fan-shaped signal can generate multiple returns – showing stronger pulses as darker and weaker pulses as lighter. Sometimes different colors are used to represent this. The compilation of a greyscale or color image can be used to reconstruct a three-dimensional image – typically of the floor of a large chamber, or the exterior view of a submerged outfall. Imaging sonar units are widely used on ROVs and AUVs.

Scanning a large area can be a time-consuming process, as even a fan-shaped beam produces a linear scan. Representing an entire area requires slow, precise movement and many scans. Multibeam sonar units alleviate this problem by operating multiple beams simultaneously to quickly scan a large undersea area. In an this context, some advanced multibeam sonar units must be calibrated for the index of refraction of a particular water sample in order to produce accurate results. Distances in pipelines are so small that calibration of this sort is unnecessary. That said, sonar heads should be regularly tested for accuracy as part of a routine maintenance program.

In most pipeline applications, any profiling sonar on the market will be suitable for debris estimates and ovality determination. Specifications should focus on the operational characteristics of the device carrying the sonar unit. What is the required travel distance between manholes? Should sonar and CCTV be captured simultaneously or on separate inspections? Is the sonar unit certified to work in pipelines of the correct diameter? How should output data be presented – movies of the sonar returns, or spreadsheets with debris locations and quantities? Avoid the temptation to specify kHz, TISCIT, or coloration schemes in order to preserve a competitive bidding process. Just as with CCTV, ask to see and approve a sample deliverable prior to commencing work on a contract.

One fundamental question remains regarding sonar inspections: why use complex sonar at all? Why not dewater the pipe for a complete CCTV inspection? This is a fair question – dewatering a line would expose all of the pipe wall to detailed scrutiny, making it possible to observe defects like cracks and corrosion that would fall below the detection threshold of sonar. The answer comes down to cost and pipeline size. Bypass pumping can be extremely costly, especially for large diameter interceptors. While it might be nice to perform a comprehensive CCTV inspection, doing so may be cost-prohibitive. In lines where significant debris has built up, crawlers may get stuck and be unable to traverse, even in dewatered conditions. In this case, floating systems will be essential to complete the inspection, and sonar is the only method to get quantitative data about debris.

In general, for large interceptors, *in situ* inspections with sonar and CCTV constitute best practices. If a sag, dropped invert, or strange eddy is discovered, emergency bypass pumping can always be requested, but it is not necessary to do so as a matter of course. For smaller trunk lines or even large feeders, the choice is left to the discretion of the project engineer. If the primary purpose of the inspection is debris quantification, the higher cost of sonar inspections may outweigh the cost of bypass pumping, especially if lower cost (higher productivity) floating systems can be employed. However, if there are any questions about structural integrity or below-flow deterioration, it may be necessary to temporarily dewater the line to truly observe the situation.

An astute observer will note that sonar is seldom included in small diameter inspection contracts (8–12″). Since these lines can be easily cleaned via flushing from the water supply, there is less of a need to do debris quantification studies. Further, even in the event of blockage, it is usually easier to jet, clean, or even dig up a failing small diameter line than it is to access a compromised section of interceptor. Finally, sonar must operate over distances greater than a set minimum in order to produce an accurate signal. For most commercially available units, this threshold is just over 2″ from the sonar head, making the technology much more applicable to medium and large diameter pipes.

LASER/LIDAR

While sonar is the only effective tool for "seeing" below the flow, there are a variety of techniques for obtaining quantitative information about the condition of a pipe in the headspace. These technologies often involve different types of laser systems and seek to complement or supplement CCTV inspections – not serve as replacements. Some types of laser systems are completely intertwined with a CCTV camera – the laser cannot exist independent of a particular camera and lens configuration.

Laser units are used to measure the exposed surface of the walls of a pipeline in areas above the flow. They work by shining lasers of different frequencies onto the pipe walls and observing the interaction between the beam and the surface (Figure 4.7). Two major technologies are in use today: visible light lasers – sometimes called structured light systems – and lidar or time of flight systems, that often use infrared lasers. While one system is not necessarily better than the other, each

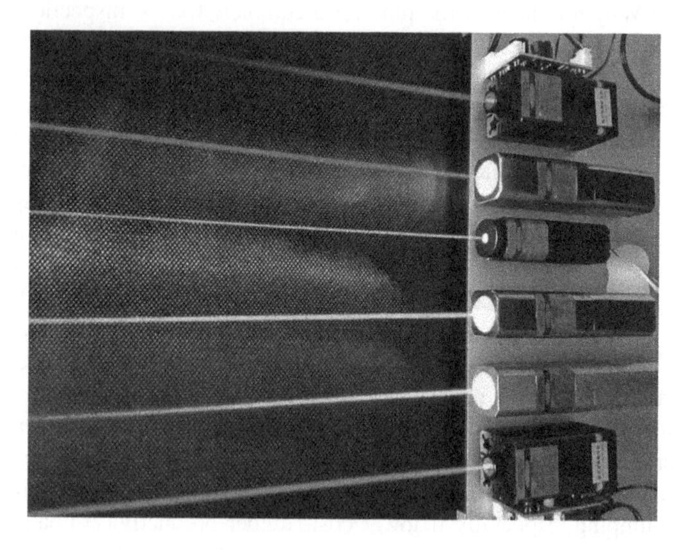

FIGURE 4.7 Lasers can generate homogenous light in a variety of colors. The stimulated emission of radiation produces a cohesive beam of light. Mirrors and prisms can be used to project this light into a number of different shapes. These specific lasers are manufactured by Q-line and feature wavelengths ranging from 405 to 660 nm. Image provided by Peng Jiajie. Creative Commons: https://commons.wikimedia.org/wiki/File:Lasers.JPG

works in a fundamentally different way and has a set of distinctive strengths and weaknesses.

Laser has become a commonly accepted noun in our vernacular; however, the term originally began as an acronym for one of the breakthrough technologies enabled by quantum mechanics. Scientists at Hughes Research Laboratories developed a way to induce electrons to transition from high to low energy states when induced – or stimulated – to do so by a photon. The electrons emit additional photons that match the stimulator in phase and wavelength, producing light that is restricted to a single frequency and polarization. This enables extremely tight focusing, rapid pulsing (as short as a femtosecond), and ultra-monochrome light production – e.g. a red laser is truly *red*, all photons within align to a single frequency of light.

These features make lasers well suited to precision measurement tasks. Compared to a standard beam of light, lasers behave in a much more uniform way. This means that a beam projected by a laser can be brighter and more focused than a typical light beam. Because all photons in a laser behave the same way, they can be precisely measured, and distances can even be computed by measuring how long it takes short pulses of a laser to travel a distance in space.

In pipelines, the two main types of lasers used to make measurements of the interior pipe walls are structured visible light and infrared lidar. Both can capture the same type of quantitative information about corrosion, wall loss, and build-up, but have very different limitations. Further, while both use light, sources of error inherent in each technology can skew results in different ways. Therefore, when using a laser system for measurement, it is critical to know which technology is being employed in order to properly evaluate it.

Laser safety can be a significant issue. Most consumer products involving lasers are relatively safe for the general public. Industrial systems, however, can use high powered lasers, and it may be necessary to use protective equipment to prevent eye damage when using laser-based devices. Most commercially available lasers do not pose a threat to users, but some discussion of laser safety and best practices is warranted.

All lasers are assigned a classification corresponding to their capacity to cause injury to users in the vicinity. These numerical classes represent increasing levels of danger and correspond to the power level of the laser beam itself. Different variations on these classes have been defined by a variety of regulatory agencies, including the American National Standards Institute (ANSI Z136), the Center for Devices and Radiological Health (CDRH) within the United States Food and Drug Administration (FDA), and the International Electrotechnical Commission (IEC). While each body defines the boundaries between laser classifications differently, the overall structure of the regimes is similar.

Class 1 lasers are considered to be generally safe. Humans can look directly at Class 1 lasers for extended periods of time and not suffer any ill consequences. Devices can be considered Class 1 through two different methods: first, the device can emit a laser beam with less than 0.039 mW of power; or second, the device must have a laser beam that is encapsulated inside a structure that prevents access during operation. Special care must be taken when repairing or maintaining these devices. Class 2 lasers are also considered safe for users – so long as the users do not stare into the beam. Exposures of less than 0.25 seconds are considered safe, and natural eye reflexes usually prevent any unintentional exposure from exceeding this limit. Class 2 lasers have beams with less than 1 mW of power. Both Class 1 and Class 2 lasers can be harmful to the eye if magnified or viewed through instruments like binoculars.

It is important to note that these laser classifications only apply to visible light lasers, as they factor in human physiological responses to light sources. Invisible lasers, such as those using infrared or ultraviolet light sources do not necessarily follow these classification schemes – even if power consumption statistics are similar. This is due to the fact that operators are more likely to face prolonged exposure to these devices, as their invisible nature prevents reflexive looking away.

Class 3 and 4 lasers pose a risk to operators and should not be used without special training and the use of protective equipment. Class 3 lasers emit beams ranging between 1 and 499 mW of power, and pose a low risk of eye damage upon unintentional exposure. Some damage is possible, however, and the risk increases as the length of exposure time also increases. Class 4 lasers have power levels of 500 mW or above and can cause eye damage immediately. These lasers can also damage the skin and should be avoided at all costs.

Most lasers used in pipeline inspection systems are derived from components that could be classified as Class 2 or Class 3 laser systems. Further, each manufacturer uses a variety of lenses and optics that can transform the beam in some way – either concentrating or diffusing it. When these devices are operated by vendor crews, the issue becomes moot, as most vendors train their crews thoroughly before operation, but when purchasing a device containing a laser for operation (laser attachments are sold for many crawlers), make sure to get documentation of how the laser is classified. If the device uses a Class 3 laser unit, and the manufacturer claims the end product falls under Class 2 guidelines, they should be able to show you why – in detail. Certain classes of lasers require the presence of a trained Laser Safety Officer during operation in order to comply with Occupational Safety and Health Administration (OSHA) guidelines. Further, even for lesser classes of lasers, understanding how to protect crews using the devices and technicians maintaining them is critical as well.

Structured light laser systems use a beam of visible light that is projected onto the interior of the pipe wall (Figure 4.8). Using conical mirrors and other projection systems, this light can be reflected into multiple points or a continuous ring. A camera system picks up images of this ring and performs computational image analysis in order to draw quantitative conclusions about the underlying data. While structured light lasers could use any type of visible light, red is the most commonly used color. Perhaps it is coincidental, but red lasers are also the cheapest available.

Early iterations of these systems projected a series of dots at various clock positions – 4-dot lasers projected spots at 12, 3, 6, and 9 o'clock. 6-dot lasers projected spots at 12, 2, 4, 6, 8, and 10 o'clock, and so on. The distances between these laser dots could be used to compute the relative diameter of a pipeline and calculate whether ovality or deflection was present. In general, more dots represented a higher quality measurement, as erroneous readings due to anomalous debris or taps could be accounted for by looking at the diameter in multiple ways. This could be maximized

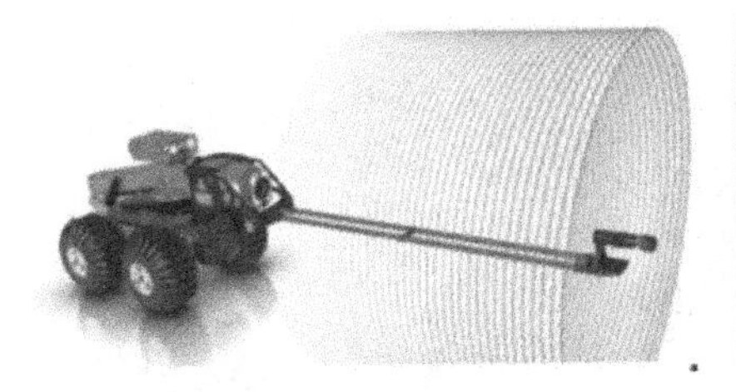

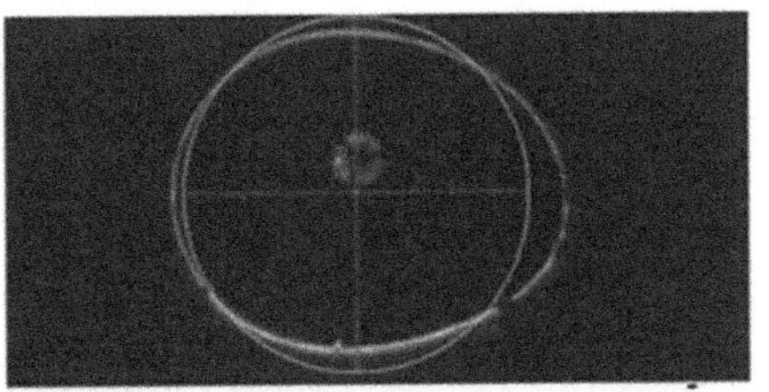

FIGURE 4.8 The left image shows a typical example of a structured light laser mounted on a crawler. Note that the laser is mounted at the end of a rigid wand. A conical mirror projects the laser 360° onto the wall of the pipe, and that projection can be seen in the image at right. This data can be used trivially for measuring ovality and can detect corrosion with proper calibration. Image taken from Cleys et al., *Pipelines.* STM (ASCE): https://ascelibrary.org/doi/pdf/10.1061/9780784480885.022

by looking at an entire ring, upon which the diameter could be measured at each pixel location – hundreds or thousands of times, giving a much more comprehensive view of deformation.

From a computational perspective, ovality measurements are trivial, as they only require counting pixels, and do not necessarily need to translate into actual inches. ASTM F1216 specifies that the percentage of ovality in a pipeline can be calculated as:

$$\text{ovality} = 100 \times (\text{mean diameter} - \text{minimum diameter}) / (\text{mean diameter})$$
$$\text{or } 100 \times (\text{maximum diameter} - \text{mean diameter}) / \text{mean diameter}$$

If a pipe has a horizontal diameter of 160 pixels, a vertical diameter of 160 pixels, and an overall mean diameter of 160 pixels, it is clear that it is perfectly round and no ovality is present. No mapping between pixels and the underlying pipe dimensions needs to occur and primitive image analysis algorithms for edge detection can draw these conclusions. Similarly, if the horizontal diameter is 160 pixels, but the vertical diameter is only 120, and the mean diameter is 140 pixels, then the ovality would be

$$100 \times (160 - 140) / 140 = 14.29\%$$

Error can occur if the robot is not centered in the pipe or is not projecting a beam that is perpendicular to the flow direction. Off-center crawlers can cause false ovality readings and typically occur when a crawler is not appropriately sized for the pipe being inspected. If the crawler is too small relative to the pipe it is easy to get off center and meander. Floating systems can also hug the walls of a pipe when transiting curves and produce erroneous readings, as well. Further, many of these lasers are mounted on wands or rods that extend forward from the crawler itself – this is necessary so that the camera can see the laser during normal operation. This creates a vulnerability, as if obstacles are encountered unexpectedly, these laser wands can become bent or misaligned. Sometimes these injuries are obvious, but sometimes they are subtle and can accumulate over time – pipeline inspection robots have difficult lives.

FIGURE 4.9 Some devices perform "multi-sensor inspection" or MSI. This image shows an example of a floating platform that has a structured light laser and CCTV camera on the top of the raft, and a sonar head mounted underneath. Note the calibration procedures that are employed to take accurate corrosion measurements. This specific device is the Profiler by RedZone Robotics, Inc. Public Domain: https://nepis.epa.gov/Exe/ZyPDF.cgi/P100C8E0.PDF?Dockey=P100C8E0.PDF

To prevent this, laser measurement systems should be regularly calibrated in non-oval pipe. A contractor should be able to show that a device was showing proper ovality calculations before and after completion of a project – with both data and raw, time-stamped video. Whenever the device is impacted or anomalous readings are suspected, the calibration should be checked to prevent errors from contaminating a large portion of the collected data. Do not agree to have this corrected in software or post processing. If the laser is mounted incorrectly, all data collected with it should be reacquired once it is fixed (Figure 4.9).

Structured light systems are also used to measure corrosion or wall loss in a pipe. This is a much more difficult process, as a detailed understanding of the interaction between the laser, mounting system, and camera optics must be obtained in order to draw these conclusions. In order to do this, the device must have a calibration generated for each operational condition that it expects to operate under. For example, if the device will be used to measure corrosion in an 18″ pipe, it should be driven through an 18″ test pipe in order to identify which pixels will be illuminated by the laser under normal operation. Expected corrosion defects need to be profiled with the laser to see which pixels correspond to various levels of corrosion or buildup.

Multiple painstaking measurements must be performed for this process to occur successfully – as many lenses induce some bending of the light, linear interpolation is not a substitute for calibration data. Further, this data must be particular to the combination of the laser, mounting system, lens, and camera. If any of those components changes, manufacturing imperfections could introduce serious errors in collected data. Similarly, bending or impact to the laser mount can create massive issues in data quality, and require a complete recalibration before use. Many times, these systems are coupled with fisheye lenses – these systems are especially nonlinear, and slight alterations in where the laser beam is visible on the camera can skew corrosion measurements by inches or more.

These systems also are inherently two dimensional. Since they produce scans or slices of the pipe at fixed locations in time, they are always anchored in reference to the robot itself. There is no true indication of the dimension corresponding to down-pipe travel or chainage. If each of these laser scans were to be stitched together, an accurate corrosion profile of the pipe could be obtained, but the pipe would always resemble a straight line. Imagine reconstructing a hallway by assembling slices that are 1 mm thick – essentially this is what the structured light laser system does. These thin slices are incapable of capturing curvature information and, thus, structured light lasers are not suitable for the creation of three-dimensional models. It is for this reason that structured light systems are sometimes referred to as 2D or “profiling” lasers.

These laser systems are available from multiple vendors, but RedZone Robotics produces many of the structured light units that are capable of corrosion analysis on the market today. These devices are integrated into RedZone’s profiler robots and are available as add-ons to various conventional crawlers. Remember, even if you purchase a profiling laser, you will likely need to maintain a good relationship with the vendor in order to ensure that calibration is frequently performed. Further, not all vendors allow end users to directly process data; some insist that data be sent to them for processing and returned as a final deliverable – at a fixed cost per foot. Units that can perform ovality assessments are available from a litany of vendors – with cost increasing with the resolution (or number of dots). Ovality calculations from measurement data can typically be performed by the end user without an additional fee.

Structured light systems claim the ability to recognize extremely fine details – small amounts of debris less than 0.5″ in size. Many of these claims are based on theoretical assumptions about camera resolution, laser beam width, and optical physics. In reality, how this data is processed can introduce additional – and significant – sources of error. For example, where an automated processing tool chooses to define the edge of a laser beam can have an impact the amount of corrosion the said position corresponds to. Moisture or fog on the lens can make a beam appear blurry, introducing an additional range of uncertainty. Finally, aligning the laser scan with the as-built model of the pipe is something that is often performed manually by a skilled technician. This can lead to significant differences in processed data from technician to technician – something that should not occur for tangible measurements of pipe parameters like corrosion.

This is exacerbated by the fact that many of these laser systems are made, certified, and calibrated by the same vendor, and are manufactured in quantities of tens. This means theoretical errors, bugs, and oversights can easily make it into production devices. While structured light is suitable for ovality measurements, some care should be taken before adopting it for corrosion condition assessment. If a vendor claims their system is able to detect a feature of a given size, a real-world test should validate these claims. Look to see if cracks and visible aggregate are apparent in laser scans, and insist on seeing data processed by more than one technician. Ask if reports are audited and by whom. A professional engineer’s stamp of approval can go a long way toward allaying concerns about data quality.

Finally, a brief note about data formats – there is no standard for structured light laser data, and unlike sonar, the information collected by this system does not always suit itself well to a tabular representation. While a table of ovalities at various chainages can be generated, the same table of corrosion values would lose some information. Would the value in the table be maximum corrosion, average corrosion, or some combination of the two? A small point of corrosion corresponding to a single piece of missing aggregate would be much less significant than a large patch at the crown of the pipe. How can this data be best represented?

An intuitive choice is the heat map or flat graph. In the heat map, the corrosion in a pipe is overlaid on a straight model of the pipe itself (remember, as a 2D measurement system, any curvature would be misleading). A color-coding scheme represents various amounts of corrosion and build-up – maybe red indicating areas with more than 2″ of material loss, orange indicating areas with 1–2″ of material loss, and yellow representing less than 1″. Shades of blue could be used to represent buildup. Cross-sectional slices can sometimes be used to show areas of concern by illustrating the estimated as-built diameter as well as the measured diameter. These visuals can give a clear sense of how much corrosion is present in a pipeline. Flat graphs take the heat maps and unroll them so that all clock angles of the pipe are visible in one "flat" image. Sometimes, sonar can be overlaid below the flow to produce a comprehensive visual of a pipe's condition (Figure 4.13).

While useful, these graphs are not well suited to detailed, quantitative analysis. Upon request, most vendors will provide raw tabular information of all the points that were used to produce the flat graph. This aggregated data can consist of millions or billions of points indexed by chainage and can result in some massive file sizes. While unwieldly, these large files contain a complete representation of the underlying data that was collected and can be fed into more complex analyses in the future.

The second type of laser system that is used for corrosion measurements is lidar. While some claim that lidar is an acronym for Light Imaging Detection and Ranging, the term was most likely coined as a portmanteau between sonar and radar. Lidar is a time of flight measurement system – small pulses of laser light are emitted, reflected off of a surface, and detected by a detector. The amount of time elapsed between emission and detection corresponds to the total distance traveled – as a function of the speed of light.

Functionally, all of this occurs in a unit known as the lidar head. The lidar head contains several key components: the laser source itself, a spinning mirror, and a detector, in addition to other supporting electronics. The laser source operates in a similar manner to the structured light lasers discussed earlier in this section, albeit with a few key distinctions. First, the light is pulsed, rather than continuously emitted. These packets of light are the fundamental unit of measurement of the lidar system, and are produced to have enough energy to be detected upon their return, yet be short enough in duration to make an accurate measurement. Further, lidar devices commonly in use today do not produce a visible beam. Instead, they operate at the infrared portion of the electromagnetic spectrum, generating light with a wavelength of either 905 or 1550 nm (Figure 4.10).

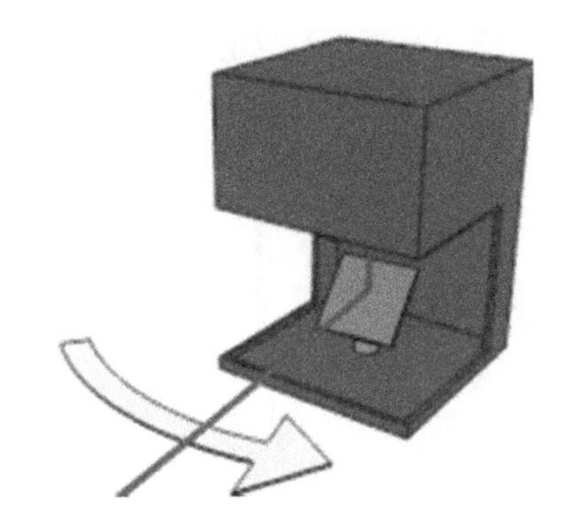

FIGURE 4.10 Lidar units are found on many ground robots and factor in to localization and mapping algorithms. In stationary devices, a laser beam is reflected off of a spinning mirror in order to sweep an area. Further rotating the lidar head itself can be used to create a three-dimensional representation. The left image showing the lidar head (yellow) mounted on a robot was provided by S. Winkvist. Creative Commons: https://commons.wikimedia.org/wiki/File:LIDAR_equipped_mobile_robot.jpg Public Domain: https://en.wikipedia.org/wiki/Lidar#/media/File:LIDAR-scanned-SICK-LMS-animation.gif

While a case could be made for using either frequency, said discussion is beyond the scope of this text. In water and wastewater pipes, almost all lidar systems use 905 nm lasers, as water is much less absorbent at this frequency. By passing through small water droplets, 905 nm light pulses are able to overcome water droplets on the lens of the lidar head and small amounts of spray or mist in a pipe. For this same reason, 905 nm devices are much more popular in self-driving car applications – the weatherproofing being essential to operation in this use case, as well (Figure 4.11). Though invisible, the 905 nm emissions of lidar units are considered to be as safe as Class 1 visible lasers.

The pulses of light are projected onto a spinning mirror which is able to send light out in a vector across a straight line in space. By moving the lidar head itself, this line can be traced multiple times in order to eventually construct a three-dimensional representation of the space being scanned. The mirror must be rotated with precision timing so that the exact angle of emission of each pulse is known – otherwise accurate image reconstruction will be impossible. The detectors that receive the reflected pulses can be classified as coherent or incoherent. Coherent systems can detect changes in phase of the reflected light, and incoherent systems can detect changes in amplitude. In most pipeline applications, this distinction is not significant, as the time in flight is the only parameter used to calculate distance traveled. Phase and amplitude differences can be used to provide further analysis such as material identification.

Unlike structured light systems, lidar heads are commercially available from a variety of vendors including Sick, Velodyne, Pepperel & Fuchs, and others. These vendors produce lidar units for industries well beyond inspection robots, ranging from autonomous vehicles to industrial automation. They produce thousands and thousands of these devices and have well-defined quality control and testing

FIGURE 4.11 An example of a point cloud of a city street produced a lidar unit mounted to a vehicle. Since each data point contains a time of flight measurement, detailed, geometrically accurate representations can be created. Coloration can be set to correspond to distance away from the unit itself, or other user-defined schemes. Image provided by Daniel Lu. Creative Commons: https://commons.wikimedia.org/wiki/File:Ouster_OS1-64_lidar_point_cloud_of_intersection_of_Folsom_and_Dore_St,_San_Francisco.png

procedures. They also have well-understood calibration and MTBF recommendations, since they have been deployed in much larger quantities. While larger companies are not always better than smaller ones, the lidar manufacturers include experts in both lasers and optics in their R&D divisions, and have engineering budgets that are higher than most vendors of sewer inspection equipment. The spec sheet for a Velodyne puck does not require the same level of scrutiny as a newly developed structured light system.

A major advantage of lidar systems over structured light is that distance measurements are automatically computed by the device itself – meaning lidar units are completely independent of any associated camera system. This means lidar heads can be easily swapped between robots without the need for lengthy and complicated calibration procedures. Further, their stock calibrations are certified by external manufacturers and can be easily tested against reference structures or patterns. Much less custom engineering is required to use lidar in an inspection robot when compared to structured light systems.

An off-the-shelf lidar unit can act as a suitable drop-in replacement for a structured light system mounted on a crawler or floating platform. By allowing the mirror to rotate as the device moves down the pipe, multiple slices can be combined to produce a straight reconstruction of the interior of the pipe – a 2D representation similar to structured light systems without the need for calibration. In order to capture bends and other downpipe information, a way of capturing the downpipe dimension is needed.

One strategy is to face the lidar head down the pipe, and rotate it about the axis of the robot. The combined rotation of the lidar head and the mirror inside enable the sweeps of the lidar beam to trace a three-dimensional slice of the pipeline. Realistically, the lidar unit can return data from more than 100′ away, but points will be most dense in the 5–20′ area in front of the robot itself. This broad swath of pipeline is sufficient to capture bend and radial changes, and creates a true three-dimensional image – that still has its origin at the robot itself, but has usable depth information.

The drawback to this technique comes from the fact that the device transporting the lidar unit must come to a complete stop in order for the rotation of the lidar unit to complete. These "stop scans" can dramatically increase the length of an inspection. For example, if each scan takes 45 seconds to complete, and must be repeated every 5′ in order to ensure full coverage of a pipeline, that would add an additional 45 minutes to an inspection. For longer runs, traffic control and flow management may make the cost of these extended inspections unpalatable.

Newer lidar units have multiple channels or laser beams operating in parallel. The Puck units by Velodyne can be purchased with either 16 or 32 channels available – spaced out over a 30° span. Mounting one of these multi-channel units statically can capture a large slice of the pipeline (5–15′) that is sufficient enough for 3D reconstruction. As these units are increasingly used in self-driving cars and autonomous vehicles, their cost continues to decrease. Expect to see more multi-channel lidar units employed in inspection robots in the years to come.

Lidar does have some significant downsides. First, it can be challenging to use lidar in small diameter pipelines. Minimum return distances for several popular Velodyne units are on the order of 1 m, meaning that the unit must be mounted in such a way that pipe walls are more than 1 m apart along any axis. While some creative mounting solutions have been employed, this can limit the use of these devices to 72″ pipelines and larger. Other units by Sick can detect objects that are 0.1 m away, making them easier to deploy in smaller pipelines (but at the expense of multiple channel collection).

Further, there is an inherent source of error present in lidar units since there is some variation in the emission of pulses. Individual data points on the Sick LMS 200 can have systematic errors of ±15 mm and random errors of ±5 mm. Cumulatively, this could cause points to be off by 20 mm from the true value. While the random or statistical errors are not of concern, the systematic errors – which are largely tied to environmental conditions – could lead to an over- or underreporting of diameter by more than 1 cm. Velodyne lidar units report accuracy of ±2–3 cm depending on the model. Just because these units can give a complete map of a pipe wall's interior does not necessarily mean that they should be trusted (Figure 4.12).

The best way to assess the validity of lidar measurements is to ensure that control data is obtained at the same time and in the same place as an actual inspection. Have the lidar unit scan a box or tube of a known size prior to commencing inspection work. Analyze this data in the same way as the inspection data and verify the actual measurement error in real-world conditions. If your 24″ box measures to exactly 24″, you can feel confident in your inspection data. If your 24″ box measures to 24.5″,

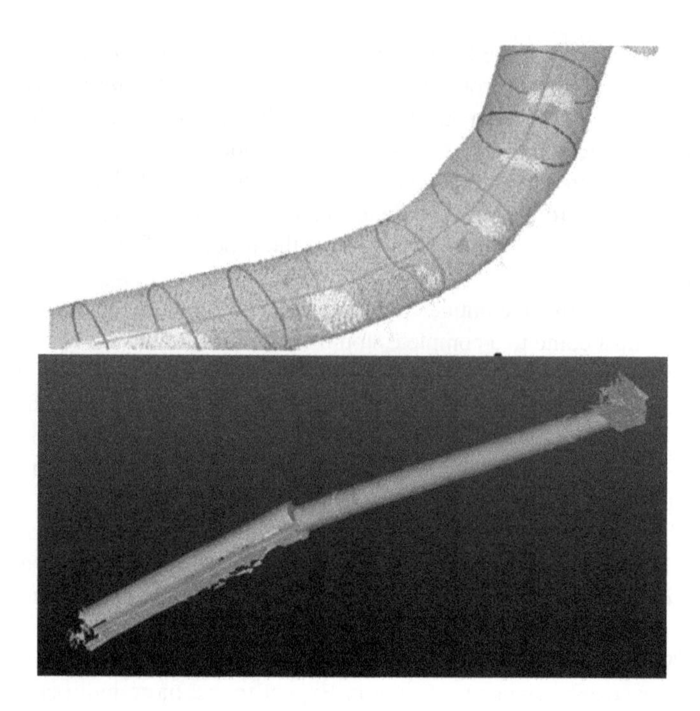

FIGURE 4.12 When rotating lidar devices are used, individual scans must be stitched together in order to produce an extended model of the pipe itself (top). Errors or uncertainty in the stitching process can lead to significant sources of error. However, when corrected, lidar can produce a detailed 3D model of a complex pipe with size, shape, and elevation changes (bottom). Image from Cleys et al., *Pipelines*. STM (ASCE): https://ascelibrary.org/doi/pdf/10.1061/9780784480885.022

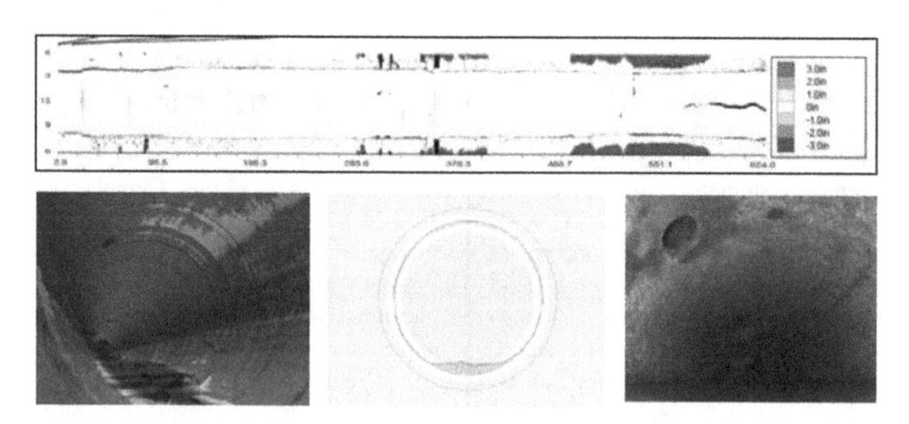

FIGURE 4.13 Laser data can be presented in a number of different formats. An example heat map is shown at top, where the interior of the pipe is unwrapped into a 2D projection. Corrosion measured by the laser is shown as areas of red, orange, and yellow. The bottom images show other possible representations. Public Domain: https://nepis.epa.gov/Exe/ZyPDF.cgi/P100C8E0.PDF?Dockey=P100C8E0.PDF

you know that there may be issues in the lidar data that was obtained. It is possible to correct these points, but only if they can be tied to a calibration object in some way – to do otherwise would be guesswork (Figure 4.13).

The published tolerances of lidar tend to scare away some potential users – a fact which is understandable. When a self-driving car must make a decision in a fraction of a second, a few centimeters of uncertainty are hardly significant in the scheme of operations. In the context of a pipeline with walls that may have a thickness of 4″, standard error on a lidar unit could represent nearly 25% of this thickness. While true, remember that this is known upfront and is published by the manufacturers of lidar sensors. Structured light manufacturers often hype accuracy and are reluctant to discuss error – just because they don't publish that information doesn't mean it doesn't exist. Always insist on viewing calibration data to verify claims.

Lidar units can also fail in unsettling ways. Just as impact can adjust the misalignment of structured light systems and necessitate a recalibration, so too can lidar units be damaged in ways that introduce error – albeit in ways that are harder to detect. Sudden impact or wear and tear over time can cause the bearings that spin the mirror inside the lidar units to rotate unevenly. This wobble in rotation causes a slight ovaling of path that makes the pulses follow a nonuniform path. In some cases, this means that pulses may travel different distances based on this offset in rotation. If the lidar head moves, gravity can cause this wobble to gradually shift from side to side, creating errors in data that are extremely difficult to decouple and correct. This problem is exacerbated in three-dimensional systems, where two different axes of rotation can fail independent of one another.

The telltale sign of this condition is a phenomenon known as double walling. If a lidar unit makes multiple passes to scan a pipe wall, it will report the location of the pipe wall in a slightly different place each time. Unfortunately, the true location of the pipe wall is not simply the average of the two ghosted images – often it is a nonuniform distance between the two. Double walling can creep into inspections and can be very difficult to detect – and can introduce serious sources of error. Ask to see the raw data produced by lidar inspections to verify that double walling is absent. Each scan should include a sharp, crisp image of the pipe wall – not two parallel cylinders. This double walling would also appear in calibration data and can be detected easily – if an inspection vendor looks for it.

The data generated by lidar systems follows an industry standard format. Each returning pulse of laser light results in a distance measurement corresponding to the angle of rotation of the mirror at emission. This generates a point in three-dimensional space. After aggregating millions of points, the result is a point cloud or a three-dimensional model of the lidar pulses reflecting off the walls of the pipe. This point cloud can be viewed in free software – such as the excellent Cloud Compare – and allows an engineer or data reviewer to place themselves inside a virtual model of the pipeline.

These points are difficult to use computationally, as they do not constitute a surface. Calculating corrosion information would require separate calculations to be performed for each of the millions of points that were collected – something that would be computationally inefficient and misleading. Instead, point clouds are often transformed into meshes – surfaces that are formed by creating small triangles from adjacent points. These triangles are stitched together to produce a contiguous surface representing the interior of the pipe. It is much easier (and more representative) to calculate distance information for surfaces generated by a mesh. While meshes can be much less attractive than point clouds, they are more frequently used for analysis and can be unrolled to produce the same heat maps and flat graphs as structured light systems. Meshes can be easily viewed via the free software program, MeshLab.

With three-dimensional lidar, it can be tempting to stitch together multiple scans in order to generate a true image of the pipeline in space – not just a straight-line composite of slices but a real representation of the bends and curves that a pipeline undergoes under the surface – enabling one to see exactly how it travels beneath roads, streets, and possibly structures. While it is theoretically possible to accomplish this task using lidar data alone, doing so is inherently risky.

Each scan represents a three-dimensional slice of the pipeline, and while it is true that scans are broad enough to contain some information about bends in the pipeline, the process of stitching scans together to create the entire segment has the potential to compounding sources of error. The process of joining multiple lidar scans involves a process called registration, which is analogous to a best-fit algorithm – it attempts to reduce the error between multiple sets of points or planes corresponding to said points. In smooth featureless pipe, multiple orientations of laser scans can fit together relatively well; if the wrong orientation is selected, the pipe can literally appear to bend the wrong way in space – turning left when it should turn right, or moving up where it should move down (perhaps in a pressurized force main).

While a trained technician provided with detailed as-built drawings can do a reasonably good job of selecting the correct orientation, this can have the consequence of having data fit the model, rather than producing an accurate representation of the nuanced bends in a pipeline. Further, if a bend measurement is off by a small amount – such as half a degree – and the pipe continues on for a thousand feet or more – the lay of the pipe could be significantly incorrect. This issue is critical, as these bend radius reports are often used to decide on pylon placement in addition to other critical issues.

This is not to say that localizing a pipe cannot be achieved with lidar-based systems. Accurate localization is possible, so long as additional sensors are utilized. Which sensors are selected becomes a function of cost. The cheapest source of supplemental data is to use an inertial measurement unit – typically a device built into the crawler itself. The IMU contains a set of gyroscopes and accelerometers and can be used to track the position of the robot over time, even when GPS signal is not available. IMUs are notoriously subject to drift – meaning they have a poor sense of forward motion, and can incorrectly estimate the length a robot has traveled. This drift is largely a function of price – inexpensive IMUs can be very prone to the issue, while devices costing five figures produce more accurate results.

Regardless, simply extracting the orientation data from the IMU can help solve this problem. If the direction of gravity in a lidar scan can be determined, the number of possible registrations or best-fits decreases significantly. If the lidar unit is able to scan a portion of the robot, this can accomplish the same goal, assuming the device is not climbing the sides of the pipeline. A true reference for which direction is "down" can go a long way toward eliminating egregious misfits in point cloud stitching. If a bend radius analysis is performed using lidar alone, ask to see the IMU data to check the fit, and consider sharing as-builts after you see the initial model to check the accuracy of the process. Since IMU data is often built-in to a robotic crawler, it should be available at no additional cost.

Other sensors can provide much more accurate results than IMU data – but at a much greater cost. Attaching a sonde to the robot can be a useful way to map its progress from the surface. Deriving from the French word for probe, sondes are acoustic devices that produce a signal that can be detected by a specially tuned receiver at the surface. These tones are sometimes within the range of human hearing – 512 Hz is common – and a technician listening in at the surface can follow the path of the robot, guided by clues such as manhole covers. For especially deep interceptors, such as those bored through rock, this may not be practical, as many sondes cannot be detected beyond a depth of 15–25 ft.

It is also possible to attach survey equipment to the robot, in order to accurately place lidar scans and build an extremely accurate model of the pipeline. Using a total station device and attaching a prism to the robot is one method to enable a precise distance measurement – mm-scale accuracy over 1,000′. Total station devices use infrared light to make these calculations and can be rented or purchased in order to produce survey-grade inspections. Calibration of these units is required, and they do require a line of sight between the robot and prism. Measuring around a bend would likely require manned entry in order to segment the bend into multiple straight line measurements. If this level of accuracy is required, expect to pay a hefty premium.

For a pipe with a single bend and accurate GPS coordinates for the beginning and ending manholes, it is reasonable to think that sources of error could be controlled using a combination of IMU and lidar data. However, for more complex layouts, such as s-curves with no intermediate manholes, it may be worth investing in additional equipment to ensure localization is performed as accurately as possible. One pylon into a sewer can cause a construction project to quickly go off track, with multiple parties looking for someone to blame.

PIPE WALL MEASUREMENTS

The goal of both lidar and structured light systems is to obtain information about the interior condition of the pipe wall so that quantitative measurements can be made. This can be critical, as reinforced concrete structures can have a similar appearance when undergoing moderate or severe corrosion. Rules of thumb do not always apply, either – just because you see aggregate and no rebar in a known reinforced concrete pipe does not mean it is safe to conclude that corrosion is moderate. It is possible that the entire rebar cage has been consumed, and the pipe is nearing an imminent failure condition.

Laser and lidar systems can obviate the need for expensive mandrel testing in new construction or pipe about to be lined. Lasers can easily detect ovality and deflection and can make such accurate measurements of a pipeline that annular space can be calculated for a variety of liner sizes, making capacity maximization and selection a much more scientific process. Some laser units can also measure detailed angular deviations, making it possible to accurately map bends under the surface. Some caution is needed in relying on these results, as inherent tolerances in measurement can produce far-reaching sources of error when propagated.

One significant source of error should be noted that does not depend on the technical nuances of lidar or structured light systems – it applies to both, equally. All laser measurement systems only give information about the current interior dimensions of a pipeline. While they may report "corrosion" or "debris" buildup, these measurements are based on assumptions. For example, ASTM B36.10 describes certain standards for steel pipelines. A pipe with a nominal size of 16″ (NPS 16) has an outer diameter of 406.04 mm. This pipe can be obtained with a variety of wall thicknesses corresponding to schedule number. Schedule 5 has a thickness of 4.19 mm. ASTM A530 gives acceptable manufacturing tolerances of +2.37 mm or −0.79mm in the outer diameter and ±12% in the wall thickness.

Assuming no deviation, the Schedule 5 NPS 16 pipe would be expected to have an interior diameter of 397.66 mm (subtract two thicknesses from the outer diameter measurement), and this would be the "expected value" when this pipe is measured. Larger diameters would indicate corrosion or wall loss and smaller diameters would indicate buildup or tuberculation. However, a pipe with an outer diameter of 408.41 mm and a wall thickness of 3.68 mm would also be acceptable per the specification, meaning that an interior diameter of 400.53 mm would also be within specification. Unfortunately, these as-built conditions are seldom recorded in detail.

If a Schedule 5 NPS 16 pipe is inspected with a laser measurement system a decade after installation, and the diameter measures to be 400 mm, does this represent corrosion of 1.17 mm – or a loss of 27% – of the pipe wall? Or is this far more moderate corrosion – less than 0.1 mm on a pipe that was constructed to the upper end of the specification? With one measurement at a point in time far removed from construction, it can be difficult to tell. A corroded steel surface looks like a corroded steel surface – the depth of that corrosion can be difficult to tell via eyeballing or visual inspection.

The same problem can exist in concrete pipe. ASTM C76 specifies a minimum wall thickness for reinforced concrete pipes of a variety of sizes. As pipe diameter increases, so does wall thickness – a pipe with an internal diameter of 18″ has a minimum wall thickness of 2″, while a larger pipe with an internal diameter of 108″ has a minimum wall thickness of 9″. Not all reinforced concrete pipes are the same, however – some pipes are rated for a higher structural load, and have higher wall thicknesses as a result. ASTM terms these Wall A, Wall B, and Wall C, with A having the least structural support and C having the most. The same 108″ pipe constructed to Wall A standards would have a minimum wall thickness of 9″ and support 3446 lbs/ft. At Wall B standards, it would have a minimum wall thickness of 10″ and support 3865 lbs/ft, and at Wall C standards, it would have a minimum wall thickness of 10.75″ and support 4160 lbs/ft.

While all of this documentation should be retained, it is not uncommon to find as-built drawings noting the installation of 60″ reinforced concrete pipe with no note about the Wall adopted. Since even moderate amounts of concrete corrosion and rebar wear can impact the structural strength, this uncertainty could lead to premature replacement – or structural failure, depending on how it is acted upon. Further, many construction specifications permit variations in the internal diameter of a pipe by ±3%. In a 60″ pipe, that means the constructed diameter could be as much as 61.8″ without any corrosion. If this pipe were inspected 10 years after installation, and the interior measured to 62″, would that indicate uniform corrosion of 1″ around, and a loss of 1/5 the total wall thickness (assuming Wall A construction), or would that indicate a pipe that is essentially in the same condition as it was constructed, corroding very slowly?

The solution to these questions is to ensure that laser data – or any other diameter measurements – do not exist in a vacuum. If a pipe were inspected with laser measurement prior to being put in service, a municipality would have a useful map of interior diameter that could serve as a baseline for future inspections. Yes, this would add cost to a new construction project, but could add significant value when amortized over the lifespan of the asset, as it would enable accurate condition assessment profiles to be generated with minimal uncertainty. Guesswork should be minimized whenever possible in lifecycle assessments.

There is a significant amount of misinformation in the industry about the strengths and weaknesses of lidar and structured light systems – most of which can be directly tied to vendor marketing material. Some concluding thoughts about which system is appropriate may be helpful. Both laser and lidar systems are suitable for ovality measurements – these calculations are straightforward, and data should be able to be turned around almost instantly, as minimal processing is required. For corrosion analysis, both systems can be effective and are likely to have similar levels of accuracy. While structured light vendors claim theoretical advantages, software and practical concerns negate most of them. Consider the overall system – if a structured light system is tied to a specific camera, make sure that camera produces acceptable image quality. It may not be possible to pair a structured light system with a 1080i PTZ camera. Lidar systems can be bolted on to anything, and image quality need not be compromised. Further, since lidar systems come from large, well-known manufacturers, accuracy limits can be referenced from public specification sheets – not vendor calculations.

That said, many lidar systems come with a significant disadvantage – cost. Many structured light systems have been heavily optimized for fast data collection. By using a fisheye camera, timed flash system, and appropriately calibrated laser, a floating platform can quickly travel through a pipeline capturing sonar, laser, and virtual PTZ inspection data at a very rapid pace. This translates to a much lower price per foot. There are tradeoffs in this data – the frame rate can be slow, the laser wand is often visible in the video data, and the limited resolution of fisheye systems (see previous chapters) can sometimes produce subpar image quality in large interceptors. However, for a quick and cheap inspection, the tradeoffs may be worth the cost savings.

Finally, for three-dimensional analysis, there is no substitute for lidar. Only multi-channel or spinning lidar systems can capture "slices" of a pipe that are wide enough to include bend or incline information. That does not mean that this lidar data is accurate enough to reconstruct those bends without supplemental information, but it does mean that structured light or stationary lidar units cannot physically perform this analysis. Again, this three-dimensional information often can only be collected with slower inspections, and comes at a cost.

What to specify then? Focus on conclusions and data, not sensors. Consider specifying a laser-based system that can measure 0.5″ of corrosion (or whatever threshold you wish to set), and specify a minimum defect size. Ask to see calibrated data, and put a test feature in a line – if that feature is picked up in the inspection, then you can feel confident that the technology performs as advertised. Avoid discussing laser wavelengths or lidar sweep rates and focus on the data that the inspection needs to generate. If you need three-dimensional information, consider specifying spinning lidar, but perform the same checks and verifications. Also give thought to a data format – are point clouds necessary for modeling, or would PDF or Excel summaries of relevant data be the most useful? Should heat maps be delivered as images or numerical data that can be imported into other systems?

5 Methods for Detecting Hydrogen Sulfide Gas

As noted previously, hydrogen sulfide is a leading threat to concrete products inside sewer systems (Figure 5.1). This can include reinforced concrete structures, cast in place pipes, junction boxes, manholes, and even the cement mortar holding together brickwork. This corrosion occurs due to the combination of multiple factors, including bacterial activity, cracking due to the expansion of the products of corrosion, such as gypsum, and decreasing pH levels that combine to produce an acidic environment in which several hydration compounds can be easily broken down and dissolved. The extent of this problem is significant – hydrogen sulfide corrosion can cause rapid deterioration of pipelines and premature failures. Some systems have observed a wall loss of 12 mm/year, while higher rates have been seen in the laboratory.

Thus, it is critical to detect and prevent this process whenever possible. Several sensors and techniques have been established to detect the level of hydrogen sulfide in pipelines, and assist engineers in developing degradation profiles. While the industry has developed a number of "rules of thumb" that can accurately predict problem areas in collection systems, these new methods can refine or enhance previous estimates which have not been shown to be comprehensive.

While hydrogen sulfide causes a significant amount of structural damage in sewers, it is more important to note that it can be both endemic in sewer systems and deadly to humans. Proper care and procedures must be observed whenever hydrogen sulfide could be present in an environment, regardless of whether or not it is suspected. High concentrations can lead to immediate immobilization and death, and sewer workers die every year from exposure. Changes in an underground environment can lead to the rapid release of concentrated pockets of gas, so never assume that a previously inspected or well-known pipeline is safe. Always take precautions (Figure 5.2).

While the mechanism of hydrogen sulfide corrosion has been discussed in earlier chapters, some information about the source of the gas is relevant to its detection. When wastewater flows through a pipeline, a slime layer forms at the interface between the flow and the wall of the pipe. It consists largely of a mixture of bacteria and solids and tends to form in all areas where wastewater flows. The name slime is appropriate – it can be very slippery. It is very easy to lose one's footing on the rounded invert of a pipe due to these slime layers – even in conditions of low flow. The thickness of the slime layer depends on the scouring ability of the wastewater flow. High velocity flows with large amounts of rough solids will tend to keep layers thin, while slow flows without scouring lead to the buildup of thicker layers – up to about 1 mm, maximum.

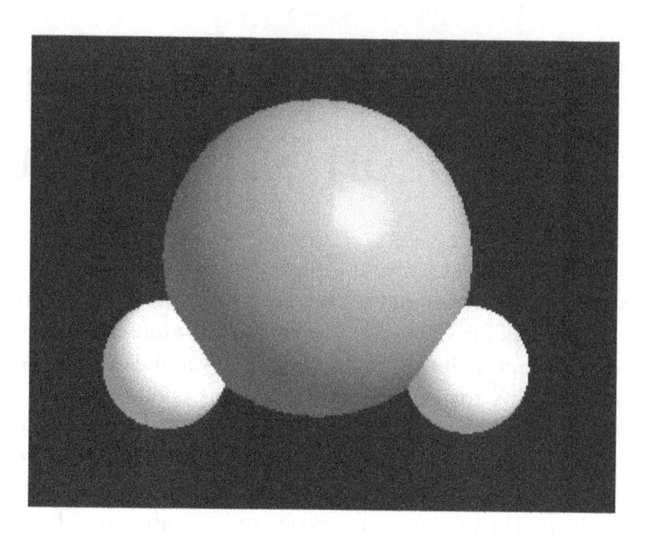

FIGURE 5.1 Space filling model of the hydrogen sulfide gas molecule. It has the same molecular structure as water, but the absence of hydrogen bonding causes it to exist in a gaseous state at room temperature. Public Domain: https://commons.wikimedia.org/wiki/File:Hydrogen_sulfide_-_Space_filling_model.png

	10 min	30 min	60 min	4 hr	8 hr
ppm					
AEGL 1	0.75	0.60	0.51	0.36	0.33
AEGL 2	41	32	27	20	17
AEGL 3	76	59	50	37	31

* Level of Odor Awareness = 0.01 ppm

FIGURE 5.2 Table of acute exposure guideline levels (AEGL) for hydrogen sulfide gas exposure. Level 1 indicates discomfort and irritation. Level 2 indicates serious side effects that can be irreversible. Level 3 indicates life-threatening side effects, including those that are fatal. Public Domain: https://www.epa.gov/aegl/hydrogen-sulfide-results-aegl-program

Note that this layer is not the same as debris buildup below the flow – this slime layer that is able to harbor the growth of bacteria is anchored to the wall of the pipe on some level and exists at the interface. Its nature prevents it from extending to larger thicknesses. Although debris may accumulate on top of the layer, they are usually not anchored and can be easily dislodged. In contrast, the slime layer can be affixed very strongly, though its thickness can vary based on environmental conditions.

This slime layer can be further broken down into three sublayers, which develop as the system matures. The outermost sublayer is filled with aerobic bacteria that oxidize sulfur compounds. The functioning of this layer is dependent on sufficient levels of dissolved oxygen in the wastewater flow, as this oxygen is consumed by the aerobic digestion mechanisms. Below this is a sublayer full of anaerobic bacteria. These anaerobic bacteria reduce sulfates in wastewater and generate hydrogen sulfide. When this anaerobic sublayer is topped with a sufficiently active aerobic sublayer, the hydrogen sulfide gas is further oxidized and no release occurs. If this layer is not sufficiently active or the dissolved oxygen content in wastewater is low, not all of the H_2S will be consumed, and some of it disperses into the sewage in an aqueous form, and ultimately can be released as a gas at the surface. The innermost sublayer is inactive or inert, as no bacteria are functioning.

Many methods of hydrogen sulfide control focus on managing the activity of these biological layers in order to minimize the release of corrosive H_2S into the environment. High-velocity flows encourage aeration and an increase in dissolved oxygen levels, while storage or areas of stagnation can easily consume dissolved oxygen, leading to pockets of gas release. High levels of nutrients can also increase bacterial activity – increasing the biochemical oxygen demand increases the rate at which dissolved oxygen is consumed within a system. Striking a balance between oxygen addition and consumption can be key to inhibiting H_2S release and the corrosion that ensues (Figure 5.3).

It is important to note that sulfur oxidizing bacteria do not only exist in the top slime sublayer within the wastewater flow. They can also exist in a moist layer at the surface of a concrete pipe – in the exposed head space. This is a critical distinction, as these bacteria will still oxidize hydrogen sulfide on contact, but will retain the sulfuric acid that is produced in the moist surface area, further reducing pH and accelerating corrosion. When this sulfuric acid is produced in the below-flow slime layer, it is dispersed in the wastewater and carried downstream to the treatment plant – it does not concentrate in sufficient concentrations to impact corrosion. Thus, the same bacteria can inhibit and encourage corrosion, depending on their location within the bacterial colony in the pipeline (Figure 5.4).

Various methods of controlling hydrogen sulfide corrosion focus on altering the relationship between oxygen demand, dissolved sulfides, and oxygen in the wastewater flow. Chemicals added to the wastewater such as hydrogen peroxide or potassium permanganate act as oxidizing agents that can facilitate the oxidation of hydrogen sulfide – complementing the behavior of the oxidizing bacteria. Other chemicals such as iron chloride or other iron salts can chemically react with aqueous hydrogen sulfide to form iron sulfide – a compound that is not water soluble and effectively traps the harmful sulfide in an inert form that can settle out of solution and can be carried along with the flow for treatment. It is also possible to raise the pH to the point where the anaerobic bacteria are unable to function by adding large amounts of a base like sodium hydroxide.

FIGURE 5.3 In addition to the health effects, hydrogen sulfide is a major contributor to microbially induced corrosion in concrete pipes. This example shows a concrete pipe that has sustained significant damage due to this process. Image taken from Teply et al., *J Pipeline Syste Eng Pract.* STM (ASCE): https://ascelibrary.org/doi/pdf/10.1061/%28ASCE%29PS.1949-1204.0000327

Other solutions are mechanical in nature – ventilating problem areas with fans or blowers and replacing the environment with fresh air or nitrogen. While this is frequently employed prior to manned entry, it can also be part of a standard maintenance program, especially in problem areas. By ensuring that the atmosphere is regularly exchanged, levels of hydrogen sulfide gas can be reduced through venting. This can have the added benefit of lowering the moisture content, preventing the growth of bacterial layers in the headspace.

Regular cleaning can also reduce hydrogen sulfide production. Debris buildup can create eddies and turbulence – reducing flow velocity and releasing dissolved gases. This can also lead to the deposition of matter which can provide nutrients for bacteria and increase the overall oxygen demand. Removing debris regularly can prevent this from occurring – and maximize available system capacity.

It is also possible to construct pipes using materials that are resistant to hydrogen sulfide corrosion if it is known that buildup will be an issue prior to construction. Concretes can be constructed with built-in antimicrobial agents, which inhibit the growth of bacteria-containing layers at the surface. Sulfur concrete can resist this attack specifically and is also resistant to abrasion, making it a sound choice for applications needing corrosion resistance. Traditional concrete structures can also be constructed with an impermeable lining that prevents microbial-induced corrosion. Typically, these layers are comprised of polymers and can be cast during construction.

Similar effects can be used to rehabilitate pipe that is already in service. The deployment of cured-in-place liners can coat a damaged concrete pipe wall with an

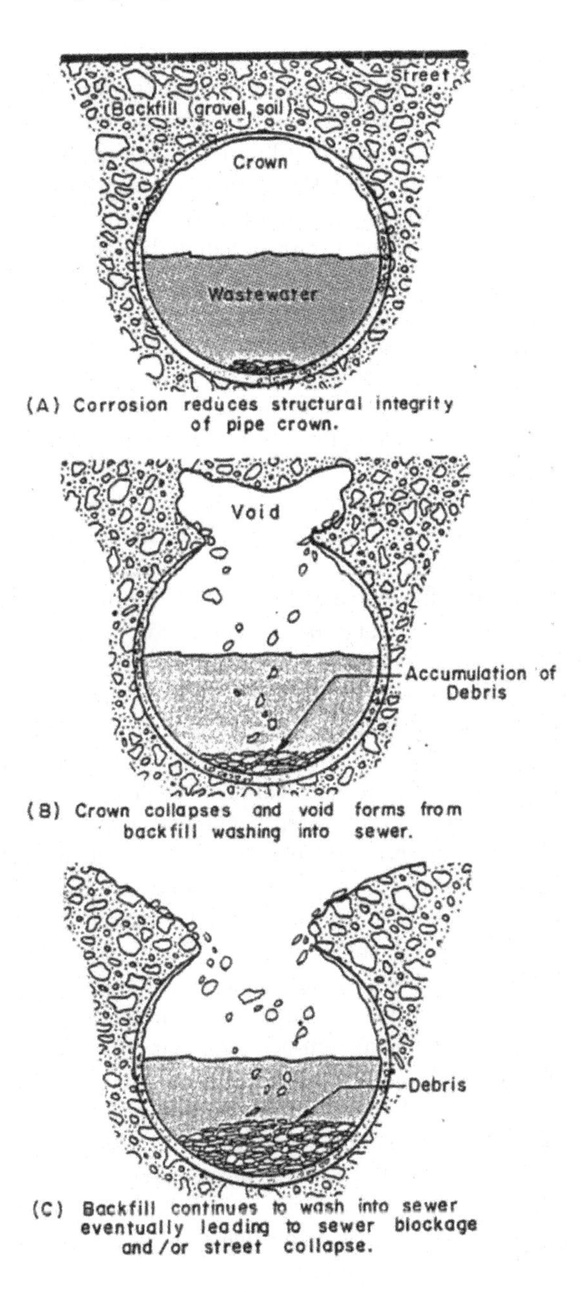

FIGURE 5.4 Illustration of the consequences of microbially induced corrosion – pipe collapse due to a loss of structural integrity is a real consequence, and can occur well before an asset has reached the end of its design life. Public Domain: https://nepis.epa.gov/Exe/ZyPDF.cgi/00000L69.PDF?Dockey=00000L69.PDF

impervious material that prevents further corrosion. Slip liners can also be used to insert a slightly smaller pipe into the host, preventing further breakdown of the concrete structure. Other texts cover pipe lining and other forms of trenchless rehabilitation, so they will not be discussed in detail in this section, but they do provide an opportunity to arrest corrosion and restore structural integrity to damaged lines – if the corrosion is caught early enough in the process.

Thus, it is clear that detecting hydrogen sulfide buildup is critical to preventing corrosion from causing advanced stages of decomposition. Long-time managers of collection systems know of several telltale methods to detect gas buildup. First, knowledge of the system can reveal locations that are prone to high levels of gas – lines that have low flow, turbulent wet wells, and other structures foster conditions that will encourage the release of gas. Proactive inspection of these areas should be part of a standard maintenance program.

Sulfur also has a characteristic odor, so areas in which residents complain of foul smells, rotten egg odors, or sewer gas are also likely areas where H_2S corrosion could be of concern. Sometimes the areas of complaint can be far removed from the actual corrosion due to the placement of manholes or other vent structures, but they provide a logical point to begin investigating the source of the gas. A CCTV inspection in the vicinity of an odor complaint can reveal visual evidence of corrosion, debris buildup or an area of stagnation that could be contributing to the production and release of hydrogen sulfide.

That said, these indirect measurements are often less than desirable. As a result, a variety of commercially available hydrogen sulfide sensing technologies have been developed. Many of these are similar to the gas monitors that employees wear when entering confined spaces. These devices will alarm when the concentration of H_2S (as well as other gases such as CO_2) reaches a predetermined threshold, giving the employee a chance to remove himself from the hazardous situation (Figure 5.5).

As an aside, it is unfortunately common for some to ignore gas meters when entering confined spaces. Anyone who has smelled hydrogen sulfide gas knows that it has a characteristic rotten-egg smell. Thus, the "logic" goes, why use a gas meter, when a whiff of rotten eggs means danger and gives ample opportunity to retreat? Unfortunately, this same thinking is what leads to inoperable gas meters sitting in the back of trucks rather than being returned for repair. While it is true that humans can smell H_2S in safe concentration ranges of a few parts per billion, the human nose is not a reliable gas detector. Exposure to concentrations of 30 ppm or greater can actually inhibit an individual's ability to smell the gas, causing said person to remain in a hazardous situation long enough for damage to occur. At these concentrations, the smell can shift to one that is more sweet, and can be accompanied by irritation of the eyes and lungs as well as nausea and dizziness.

While 100 ppm has been recognized at the threshold that poses an immediate danger to life and health (IDLH), exposure to lower amounts of H_2S has been shown to cause significant side effects, especially if said exposure is chronic or over long periods of time. Some studies have shown the development of nasal lesions after prolonged exposure to hydrogen sulfide with a concentration of 30 ppm, and others have shown the development of neurophysiological abnormalities with chronic exposure to 100 ppb H_2S. Regardless of the specific thresholds, it is clear that hydrogen

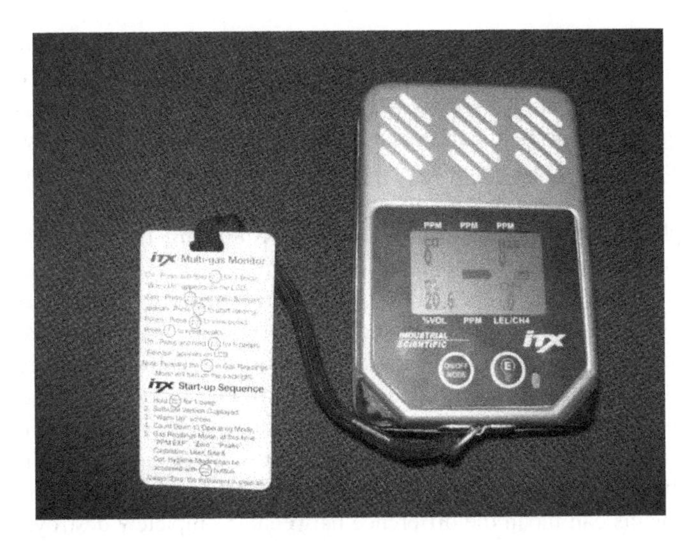

FIGURE 5.5 Gas sensor that is capable of detecting O_2, CO, H_2S, and the lower explosive limit of CH_4. This particular sensor is produced by Industrial Scientific, and the photo was taken by Sansumaria. Creative Commons: https://commons.wikimedia.org/wiki/File:Detector_for_Gas.jpg

sulfide should not be taken lightly, and meters should always be worn in environments where the gas may be encountered.

Most meters operate through one of three mechanisms – solid-state (or metal oxide semiconductor – MOS) sensors, electrochemical sensors, or optical sensors. Each class of sensors operates according to a different physical principle and has associated strengths and weaknesses. Optical sensors are least commonly used in wastewater settings. More frequently found in industrial and hydrocarbon settings, these sensors shine a beam of light through a sample of gas onto a photodetector. The wavelength of the light is selected so that it will be selectively absorbed by H_2S. The higher the concentration of hydrogen sulfide, the more absorption that takes place – the less light received by the photodetector.

Optical sensing systems do not generally have consumable parts and can function for extended periods of time with minimal maintenance. Light sources and photodetectors can experience physical failures, but such events are rare and unusual. The reason why these devices are uncommon in wastewater settings is that they require an extended operational range to function. These open path detectors send light over an extended range to the photodetector and monitor a large range of space. Further, while the presence of hydrogen sulfide may be detected, absorption will not always translate cleanly to a concentration. Fixed point detectors can mitigate this to some extent, but require careful consideration of placement. Some optical detectors will channel gas through a flow cell for testing, avoiding direct contact with the light source or photodetector – a tactic which can increase longevity.

Electrochemical systems are some of the most common devices used in field operation. These units contain two main components – electrodes and an electrolyte.

Gas is able to diffuse into the device through a semipermeable membrane. When a hydrogen sulfide gas molecule comes in contact with the electrode, a reaction occurs and a change in current is observed. A calibrated table of values associates the electrical changes with the chemical concentrations. Different types of gases will require different sensor materials – some of which may actually be consumptive, meaning the process of detecting the gas eats away at the sensor to some degree, requiring regular replacement of components. This is most commonly true for oxygen sensors wherein lead is converted to lead oxide.

The vast majority of hydrogen sulfide sensors are nonconsumptive and do not have these issues. However, meters that detect the presence of multiple gases may have a mixture of sensors – check the manufacturer's documentation to ensure that calibration and maintenance standards are adhered to. Further, just because hydrogen sulfide sensors are nonconsumptive, does not mean that they will last indefinitely. Damage, contamination, and leakage can all cause a sensor to read erroneously. Further, since the electrolyte used in the reaction is often a type of acid, rapid attention to problems can mean the difference between a completely destroyed unit and a minor repair.

Electrochemical sensors have very fine sensitivities and make repeatable measurements. They do require regular calibration to account for buildup and changing chemical properties at the electrodes. While each manufacturer prescribes a different calibration schedule – six months is a common interval – challenging environmental conditions may require more frequent calibration. These sensors also have relatively quick response times, though some delay is present. Response times are measured in terms of T-values, which correspond to a percentage of full-scale measurement values. For example, if a sensor has a T30 value of 15 seconds, that means it takes the sensor 15 seconds after exposure to reach 30% of the full-scale measurement value. Typical T90 values for electrochemical sensors are on the order of 30 seconds – fast, but not instantaneous. This means when a rising H_2S concentration is detected, best practice is to retreat and evaluate, rather than remain *in situ* to observe changes. A toxic pocket of gas could quickly immobilize someone before the meter shows the extent of the concentration. As soon as gas is detected, move to safety and assess.

The downside to these sensors is that they are not suitable for a wide range of environmental conditions. Extremely cold environments cause serious delays in the response times of the sensing elements, and hot, dry climates can cause the electrolyte to evaporate. Further, these sensors can be sensitive to other gases, so it is important to note this cross-sensitivity in unusual environments – such as those with industrial process controls. In the wastewater ecosystem, this phenomenon is generally well understood and is not usually a concern.

Finally, solid-state or MOS sensors feature a thin, resistive film at the core of the sensing unit. Each gas molecule coming in contact with the film produces a small but measurable change in electrical conductivity. Amplifiers and other transistors can increase this signal to the level where it can be usefully shown on a display or converted to an audible tone. These films can last for extended periods of time and operate under a range of temperature conditions – a significant

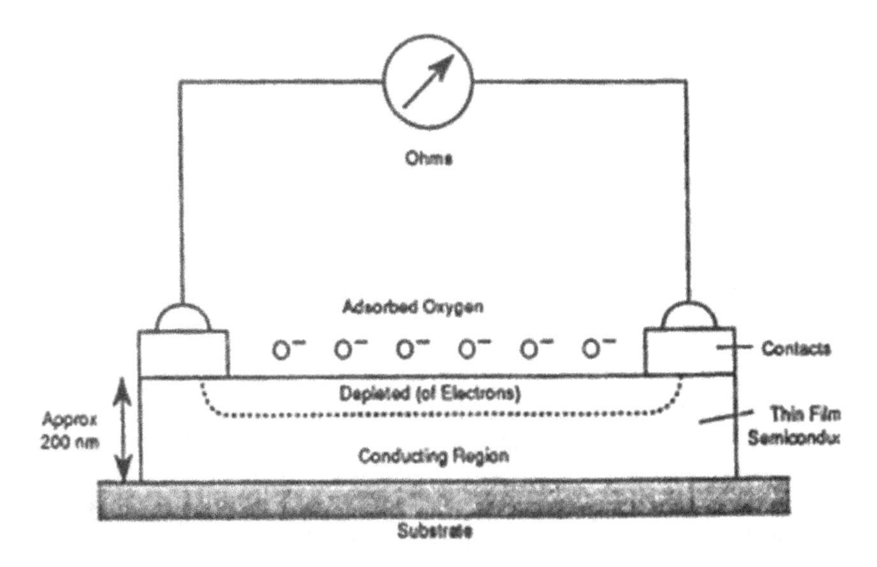

FIGURE 5.6 Schematic illustration of the structure and operation of a solid-state gas sensor. This example is showing the detection of oxygen, but similar principles could be used to detect other gas molecules. Public Domain: https://www.govinfo.gov/content/pkg/GOVPUB-C13-98816b4e8ec5fc8cf75e0f48b7d612c0/pdf/GOVPUB-C13-98816b4e8ec5fc8cf75e0f48b7d612c0.pdf

advantage over electrochemical units. They can be impacted by changing oxygen levels in the environment, and they can have very slow response times. Some newer sensors use nanostructuring to improve this significantly – to the point where they can respond more quickly than electrochemical units, but not all devices are alike (Figure 5.6).

The troubling part of MOS sensors is the fact that they can have a tendency to fall asleep, without notification. This same idea dates back to the literal canaries in coal mines. When birds would fall asleep, they could not be counted on to detect toxic conditions and alert workers – so too with gas sensors. The solution with canaries was to "bump" the cages to ensure the birds were alive and awake. The name has carried through to modern electronic devices, and "bump testing" refers to exposing the sensor to a high level of H_2S until it alarms in order to "wake up" the device. Bump testing is not a replacement for calibration – it does not show that the sensor is able to detect specific levels of hydrogen sulfide. Instead, bump testing just shows that the device can and will respond to dangerous levels of the gas. Samples for bump testing can be purchased in aerosol cans, balloons, and other gas storage devices.

Best practice calls for performing a bump test prior to each sensor use. While bump testing may not be required for all types of sensors (electrochemical sensors do not fall asleep in the same way as MOS sensors), it has the benefit of functioning as a full-system check of any sensor that is focused on results. Regardless of technology, if a sensor cannot detect a dangerous concentration of hydrogen sulfide, it should not be relied on, and is in need of maintenance or repair.

DEPLOYMENT CONSIDERATIONS

The previous overview of hydrogen sulfide and common gas sensing techniques is missing a key practical element. While electronic sensors may be able to accurately detect hydrogen sulfide, how should they be deployed in a wastewater collection system in order to obtain the best results? While it is obvious that workers should wear meters and alarms when performing tasks that have the potential to come in contact with hydrogen sulfide, how can devices be used in ways that are not attached to individuals?

One of the simplest approaches is to modify meters and turn them into data loggers. These data loggers often resemble meters in physical packaging and may even have the same display, but contain internal storage of some kind. This memory could take the form of NAND chips soldered onto the main board of the system, or a removable flash memory card, such as a micro SD or CompactFlash unit. These devices sense H_2S in the same way as standalone meters, but save readings at fixed time intervals to show how hydrogen sulfide concentrations change over time.

Logging hydrogen sulfide concentration over an extended period of time can provide a useful way of observing real-world system behavior. Whenever a manhole lid is opened in order to access a wastewater collection system, the aerodynamic properties of the entire system change – low pressure areas open up, ventilation increases, and any gas measurement taken at that moment in time would not reflect the longer-term behavior of the system. In a manner that recalls Heisenberg's Uncertainty Principle, the act of observing the system inherently changes the system.

Installing a logger and closing the manhole lid enables gas concentrations to be detected and logged for days or weeks. Spikes in concentration may be noted, and this could be correlated with heavy amounts of infiltration. Perhaps incoming fertilizer from runoff provides enough nutrients to make bacteria more active, increasing the oxygen demand of the system. More importantly, the long-term H_2S concentrations and patterns can be observed, without being impacted by human activity or the disruption that comes from opening and closing manholes.

Logging devices are battery powered, and there is an inherent tradeoff between battery life and measurement frequency. The act of writing each gas reading to memory takes a small amount of energy. Thousands or millions of readings can easily deplete a battery. The question becomes over what time period should these readings be collected? Would it be better to collect one reading every minute and have the logger operate for 24 hours, or would it be better to collect a reading every other minute and collect data for two full days of operation? The answer depends on the specific goals of an analysis program – maybe a 24-hour period is enough to get an understanding of the system's behavior, or maybe a week of readings will be needed to understand the effect a drawdown at the wet well has on the system.

Installing meters in manholes can be useful, but the closer meters can be to the actual pipe itself, the more relevant their readings will be for understanding corrosion profiles. While hydrogen sulfide can accumulate in chambers and manholes (with visible consequences), it can also move with the flow and travel downstream in a line. A meter installed at the top of a manhole far from the crown of the pipe and the flow below may miss gas activity since it is not in proximity. Caution must be observed, as contact with water tends to short out the electrodes in electrochemical gas meters and render them inoperable. In lines prone to surcharging, caution should be taken to locate meters far from high flow levels in order to protect the units. Aggressive flow can also dislodge the meters from even secure mounting points, causing them to become washed downstream and physically lost – until they end up in a trash screen or other device.

Some of these loggers are standalone devices – the OdaLogger is particularly popular in wastewater – and require manual retrieval in order to view the collected data. Similar devices include the Data Logger by Gastec. Other devices have built-in wireless modems so that H_2S concentrations can be remotely monitored. With units such as the Logaz by Injinus, the effect of iron salts or other hydrogen sulfide inhibitors can be seen in near real time. This can allow for the development of precise and accurate dosing schedules to manage gas levels throughout the day or week. Further, custom software can automate this to some level – high hydrogen gas levels can trigger a chemical dose, creating a negative feedback system that requires minimal user attention.

Many of these systems also log the temperature of the gas at the same time as the hydrogen sulfide concentration measurement. This temperature data can be useful for developing corrosion models or profiles as both bacteria and corrosion tend to increase in activity with increasing temperatures. While a high level of hydrogen sulfide may be concerning, a high level of hydrogen sulfide paired with a warm temperature may indicate that a pipe is degrading much more quickly than predicted.

Some newer products actually measure the level of sulfides in the wastewater stream. The SulfiLogger is an example of one of these devices. Produced by Unisense, the SulfiLogger detects dissolved hydrogen sulfide gas when it is still in its aqueous form – e.g. still in the wastewater flow. These sensors could be connected via serial cable to an external logging platform or remote modem, and return much more precise measurements of sulfide concentration than gaseous sensors. A recent study in Denmark suggested that gaseous hydrogen sulfide measurements were too dependent on ventilation conditions to be useful. Measurements of sulfides in the flow correlated strongly to observed corrosion and were much more consistent over time. These in-flow sensors are not yet widely used, but their development is worth observing.

Many developers of gas sensors store data in extremely efficient, binary formats. These formats are typically used because of the small file sizes that they generate, which can tie to power consumption in some ways. The downside is that many require the operation of a proprietary software tool to view and extract this data. While most manufacturers include a license for the data viewer/converter with the purchase of a sensor, these proprietary file formats can lead to some headaches when integrating these sensors into more complex systems.

First, because most loggers are self-contained, it is difficult to integrate them into larger devices – electronically and physically. Environments requiring the use of multiple gas sensors will often physically bundle an array of sensors together and retrieve a half dozen SD cards when the collection is complete. Integrating the loggers together would require a significant amount of reverse engineering, and is beyond the capabilities of many electronically-savvy end users. Second, if anything goes wrong in collection, it can be difficult to recover data from a partial inspection. Unlike plain text file formats, where data can be viewed anywhere in the file – even if that file gets interrupted while it is being written to disk – many binary files contain information necessary for decoding at the end of the file, meaning a logging device that unexpectedly loses power 23 hours into a 24-hour deployment may have nothing to show for the effort.

Some traditional inspectors of wastewater collection systems offer to include gas and temperature inspection data along with a traditional CCTV or laser/sonar inspection, and if a sewer is going to be inspected anyway, why not collect the additional data? Almost all of these systems work by attaching a standalone hydrogen sulfide gas sensor to a robot or crawler, and powering on the device for the duration of the inspection. Most vendors of wastewater inspection equipment do not produce their own gas sensors. This is generally a good thing, as designing and constructing gas sensors involves deep knowledge of electronics, chemistry, and even nanomaterials. The sensing ranges and capabilities of each sensor can be taken from a manufacturer's data sheet, similar to lidar.

The downside is that these sensors are not integrated into the larger inspection platform – meaning they still log data as a function of time, while every other piece of data collected by an inspection crawler is indexed to chainage. Further, there are often manual steps required to switch on and activate the gas logger, as it is not built-in to the crawler itself. A team of data processors must correlate the operational time to the location of the robot in order to produce data showing hydrogen sulfide concentration, as the robot traverses a pipeline as a function of distance.

None of this is inherently bad, but it does mean that gas and temperature data has a higher risk of manual error – so some data loss is not unusual. It is also common for gas data to be lost when a logger gets wet – crawlers can be submerged, floats can flip, and while other sensors tend to start working again when the liquid flows off, gas sensors are often rendered useless for the remainder of the inspection. Some data must also be discarded in order to show gas data as a function of chainage – if the robot stops at a location for a fixed amount of time – perhaps while the camera height is adjusted, the hydrogen sulfide concentration may change significantly during this period. However, in a table of gas data vs. chainage, only one reading will be reported. Ask a potential contractor what they do in these scenarios – is data averaged? Is the most recent result used? Each should be able to speak to the algorithm employed.

It is also worth noting that gas and temperature data is typically collected at a per foot cost – meaning longer inspections will have a higher price tag, even though the act of powering on and attaching the sensor remains the same. While it is tempting to collect this data at the same time as other inspection data – is it worth the cost? At $0.50/ft, it would cost $2000 to collect gas data for a 4000 ft inspection. Meters could be rented and installed in manholes for a fraction of the cost – and operated for a longer period of time, as well. This is not to say that gas and temperature data should not be part of a multi-sensor inspection, but only to note that there may be more cost-effective ways of collecting said data.

When renting or purchasing loggers, it is critical to follow the manufacturer's requirements for calibration and maintenance – going well beyond bump testing. Hold contractors to the same standard. Any reputable contractor should be able to show when the gas loggers used for an inspection were last calibrated, and should have no problem sharing information about the original equipment manufacturer (OEM) – including links to datasheets. If gas data is lost on an inspection – say the logger comes into contact with the flow – what outcome is necessary for success? Is it okay to have a gap in the data, or does the entire line need to be reinspected? While tedious, this information should be a part of contract specifications, and ought to be established upfront, before an inspection begins. Reasonable people can disagree about the level of remediation that makes sense, and proactive documentation can resolve many contract disputes.

DEVELOPMENT OF CORROSION MODELS

Many collection system owners would like to establish a degradation or corrosion profile for lines suffering from hydrogen sulfide corrosion – how much wall loss is occurring each year, and what does this mean for structural integrity and a replacement schedule? While the engineering literature contains case studies and models for a variety of systems around the world, there is no substitute for developing a model based on local conditions. This isn't simply a matter of pride – rather the number of variables ranging from concrete composition to flow temperature to odor treatment programs can cause corrosion rates to vary significantly from location to location.

Combining hydrogen sulfide measurements with other data sources can be key to developing these models. Specifically, hydrogen gas concentration and temperature data can be correlated with multiple corrosion measurements made over time. If areas with an average H_2S concentration of 5 ppm are inspected two years in a row, and 5 mm of wall loss is seen from the first year to the second, this observation can be extrapolated to other areas of the system with similar hydrogen sulfide concentrations in order to predict corrosion rates, so long as similar construction was employed.

It may seem logical to use laser/lidar systems to make these corrosion measurements, and while the tools may be useful to some degree, it is critical to note that the products of corrosion tend to expand on the surface of a pipe wall and remain in place until they are dislodged by a force. Thus, the crown of a pipe may show a slight buildup in laser scan data. However, when that buildup is scraped by a rod or measuring stick, it crumbles away and reveals a significant amount of corrosion behind. This porous layer is especially insidious, as its porosity allows corrosion to occur behind the layer itself, and it is able to trap moisture, accelerating the rate of activity. This phenomenon can be alarming – scraping against the wall of a pipe can quickly cause more than an inch of this built-up layer to crumble away, revealing a pipe that may significantly be more corroded than was expected.

The depth of this corrosion product does not necessarily correspond to wall loss in a one-to-one ratio. That is, just because an inch of material crumbles away does not mean that an inch of formerly structural concrete has been lost. Since these corrosion products expand during formation, the ratio of corrosion product to wall loss can be much greater than unity. This also means that measuring this layer is not an accurate method of determining the amount of corrosion – which is unfortunate, as it can be easily measured through probing or scraping. It also means that a laser or lidar system will not be able to measure corrosion in this scenario.

Instead, consider using an expandable rod to measure the inside diameter of the sewer. Deploying this rod with enough force will enable the rod to push through the corrosion product. Some amount of surface preparation may make this easier. Alternatively, core samples of the pipe may be taken. While destructive, this technique provides an extremely accurate measurement of wall thickness. Intelligent use of core samples in various parts of the system can be made to convert lidar measurements of interior diameter into actual wall loss – so these may add value beyond corrosion rate calculations.

Some have experimented with the use of pipe penetrating radar at the crown of a sewer to measure the distance to a rebar cage. This technology, while intriguing, has not found widespread use in corrosion measurements – in large part due to the fact that the radar antenna must be mounted in close proximity to the pipe wall. This proximity tends to be so close that the layer of corrosion products gets scraped off in the course of the inspection, easily facilitating traditional measurement techniques.

Regardless of the method that is used, the key to developing accurate corrosion profiles is making sufficient measurements over time in order to develop an accurate corrosion rate. Measurements must be spread out enough so that corrosion has progressed notably in the interval between them. For example, measuring corrosion in a pipe two days in a row would not yield useful data, as the amount of corrosion taking place in 24 hours is so small that it cannot be practically observed by conventional techniques. Making measurements over the course of multiple years can control for seasonal variations, and can tie nicely to an annual inspection program.

Once these rates are established and correlated with hydrogen sulfide gas levels, statistical models can be used to project these findings onto the entire system. This can be used to provide much more detailed justifications for odor control programs. If adding iron salts to reduce the hydrogen sulfide concentration from 5 to less than 1 ppm will add more than five years of life to a critical interceptor, it is clear that the control more than pays for itself.

6 Leak Detection – Static Sensors and Acoustic Inspections

The inspection technologies discussed in previous chapters – CCTV, sonar, laser, gas – are all focused on describing the conditions encountered in pipelines. This should not come as a surprise in a book focusing on condition assessment. However, there is another class of inspection technologies that focuses on describing and quantifying the results of these defects. Specifically, leak detection technologies attempt to identify and estimate the magnitude of pipe defects that cause inflow and infiltration or the loss of system capacity. Many of these technologies are well suited to pressurized lines, finding adoption in water transmission systems and force mains. Other methods can be applied to partially charged systems, and can be used to identify leaks in gravity sewers, as well.

These technologies can be classified based on the sensing mechanism employed and include gaseous tracers, focused electrodes, acoustic technologies, radar systems, infrared and multispectral imaging, electromagnetic anomaly detection, and other novel methods. Each of these approaches has a variety of strengths and weaknesses and can be applied most effectively in a specific set of conditions. Further, many are dependent on manufacturer claims, and do not have peer-reviewed data to support these findings. While this is also true for some of the techniques mentioned in previous chapters, some leak detection methods are purely quantitative and do not have a visual element. That can make it difficult to intuitively tell if sample data passes the "sniff test" of a trained water and wastewater professional.

When imagining a gravity sewer, one tends to picture a series of bell and spigot pipes loosely coupled together. If these systems were truly sealed, roots would never find a way in (and public works directors would celebrate). We understand that these systems experience some amount of flow loss, but this tends to be insignificant in normal operation – backups and head pressure can expose problems at joints, but under normal flow conditions, most of the wastewater stream stays confined to the collection system.

Pressurized systems experience much more significant losses. Typical public water supply lines can have pressures in excess of 80 psi, though regulators at the demarcation point in a house can reduce this to more manageable levels – 50 psi is a common set point. Transmission mains can have pressures in the 120–180 psi range and are built to withstand surges in excess of 300 psi. At these pressures, even small imperfections can quickly cause massive losses of water. A small joint leak in a 12″ main at 80 psi can cause losses of more than 25 gallons per minute (gpm) – or more than 13 million gallons per year. These leaks can rapidly erode the surrounding soil and can cause small structural problems to deteriorate quickly (Figure 6.1).

When amassed for an entire distribution system, these losses are significant. The average water system loses approximately 16% of treated water somewhere in the distribution system. That's more than $\frac{1}{7}$ of the total amount, or the equivalent of discarding all of the treated water one day of the week. Some systems with older infrastructure report losses of more than 30%. This can cost municipal governments and water authorities millions of dollars in losses alone, and result in unexpected, embarrassing failures when sinkholes swallow traffic, or water main failures create impromptu geysers, shutting down busy city streets (Figure 6.2).

Thus, leaks are a way of life in pressurized water systems, and new technologies promise to transform what was previously a game of whack-a-mole – patching up old leaks as new ones formed – into a systemic method for municipal savings, by strategically eliminating leaks, replacing aging service lines, and reducing the need for emergency repairs (which can be 4–5 times more costly than the same procedure under normal, preplanned conditions).

Before jumping into a discussion of the various techniques for detecting leaks, it is helpful to consider why they are necessary in the first place, by considering a comparison to the status quo: visual inspection. First, in order to visually inspect a

FIGURE 6.1 While water leaks are typically quantified in measures of nonrevenue water, remember that a catastrophic failure of a main can cause significant property damage. This is not a new problem – the above image shows an intersection in Seattle in the aftermath of a water main break in 1918. Public Domain: https://commons.wikimedia.org/wiki/File:Seattle_-_6th_%26_Weller_water_main_break,_1918_(22805457478).jpg

FIGURE 6.2 When water leaks are buried below the surface, it can be difficult to visualize the scope of losses. When breaks erupt, the magnitude of losing tens or hundreds of gallons per minute becomes clear. The photo of this break in Possil Park, UK, was provided by Chris Upson. Creative Commons: https://commons.wikimedia.org/wiki/File:Burst_Water_Main_-_geograph.org.uk_-_49981.jpg

pressurized line, it is almost always necessary to dewater the pipe. Even drinking water lines do not have the clarity one might expect, especially when a crawler is kicking up debris and scale. Second, how can critical and superficial defects be differentiated? For example, if surface cracks are seen in the interior of a water line, how can one determine if these are superficial imperfections – possibly confined to an oxide layer – or deep crevices facilitating rapid discharge and water loss. Even the best lidar system cannot measure the depth of a multidimensional crack as it winds its way through a material.

Further, the limitations of inspection devices themselves tend to be unacceptable for pressurized systems. Even in dewatered systems, some pockets of residual water can remain in the invert. Typical CCTV crawlers have a bias against detecting flaws in this area. Sometimes, they can be covered up by said water pockets, but some crawlers have the tracks or wheels visible at the bottom of the video frame. This can be a significant aid to navigation when traversing debris or other challenging pipe conditions, but it means flaws in the invert are harder to spot and focus on. In pressurized lines, flaws can occur at any clock position, and this blindness is a significant shortcoming of visual inspection techniques.

Another significant distinction occurs in the form of elevation changes. While gravity sewers have a gradual, constant slope, pressurized lines can have significant changes in elevation, sharp bends, and steep grades. Tethered systems can struggle to navigate these changes and the slimy scale left behind on the interior of a dewatered line can make it difficult for crawlers to climb even slight grades. For these reasons, visual, crawler-based inspection is not the tool of choice for pressurized line inspection.

METERS AND INSTRUMENT READINGS

Perhaps the most straightforward approach to leak detection is simply to use existing instrumentation and forego the need for an additional inspection. In many water transmission and distribution systems, pressure and flow meters are installed at regular locations – including customer premises. A sudden drop in pressure or an increase in flow can indicate the presence of a leak before or in front of the meter, respectively.

However, the spacing of these sensors may not be conducive to actually finding the leak in question. For example, if several customers report a drop in water pressure, and the main has fittings to connect gauges, actually isolating the location of the leak can be a time-consuming and labor-intensive process. Some locations have gauges permanently attached in maintenance structures located at strategic locations, such as large reducer valves, pump stations, or pressure vessels – but these gauges can still require manual access for readings.

Some municipalities are fortunate to have supervisory control and data acquisition (SCADA) or SCADA-like systems that have these sensors connected to a network infrastructure that enables users at a centralized location to monitor flow control structures or devices at the periphery using digital technologies (Figure 6.3). Some of these systems even support remote actuation. However, SCADA systems have traditionally required extensive physical infrastructure and significant spend to

FIGURE 6.3 Metering is a critical part of reducing losses, and it is not necessary to undertake a full conversion to smart technology all at once. Some meters, such as the Aquadis+ shown in this photo, are designed to be compatible with Cyble smart technology – making it possible to add remote reading after installation. Photo provided by Andy Mabbett. Creative Commons: https://commons.wikimedia.org/wiki/File:Water_meter_-_Andy_Mabbett_-_2015-04-14.JPG

implement – thus, these systems are not as widespread as might be expected. Every fire hydrant would not be connected to a node on a SCADA system, as doing so would be cost-prohibitive.

Advances in technology have made it much easier to add distributed sensing throughout a water system. Low-power cellular modems enable efficient data transfer and have led to the creation of sensors that can be powered by long-life (years) batteries or even solar panels. Development of solid-state components and integrated gateways have led to much more reliable sensors that can connect to distributed hubs for organized data transmission and reduced electronic costs. These advances are part of a larger movement known as the Internet of Things (IoT).

The IoT is a phenomenon that has arisen due to the low cost of connected sensors and the high value of data. When a SCADA build out was projected to cost tens of millions of dollars, it was easy to think of connected sensors as high priced novelties of limited utility. Now, when sensors can be added for a few hundred dollars or less, and the advantages of connected systems can dramatically reduce maintenance time and labor costs, a thriving sector has arisen to support adding these sensors on to conventional systems (Figure 6.4).

Systems such as Trimble's Telog units can connect to existing sensors or meters and can even be attached to existing fire hydrants, in order to quickly create a detailed map of pressures throughout a water system. These units use cellular connections and can stream data in real time or call a centralized server at predetermined intervals in order to conserve battery life. Stevens Water, Nighthawk controls, and other vendors produce similar systems that can work with a variety of sensors. Analog and digital sensors can be quickly interfaced, and existing industry standards such as SDI-12 (1200 baud serial interfaces) are supported by most vendors. This means that it is possible to put together a system using sensors and transmission units from multiple vendors – making it easy to drive down costs. However, this can push some of the support burden on to the end user – something that should be carefully considered when calculating the total cost of ownership.

While many water system operators may install distributed networks of sensors as troubleshooting tools, the value of data makes it easy to further optimize system performance. Sensors can be set to trigger alerts under high- or low-pressure conditions, making it possible to reduce line breaks and quickly discover leaks. Further, parameters of pumping stations can be set and adjusted while monitoring the real-world impact on the system. This makes it easy to see how small adjustments impact efficiency and quickly reduce nonrevenue water consumption.

Utility customers considering these systems should be aware of the fact that increased access to data can quickly generate value – and cloud-based sensors generate a lot of data. Vendors will often include software packages and training programs to help existing personnel get up to speed – NightHawk offers iHydrant, and Trimble has their own Utility software. However, it is often worth expanding a team to bring on board one or more individuals with a background in computer engineering, databases, or data science in order to make the most of this data. Many municipalities new to these systems seek to quickly replicate existing workflows using digital technologies. While admirable, a side effect of this vision is often to discard data

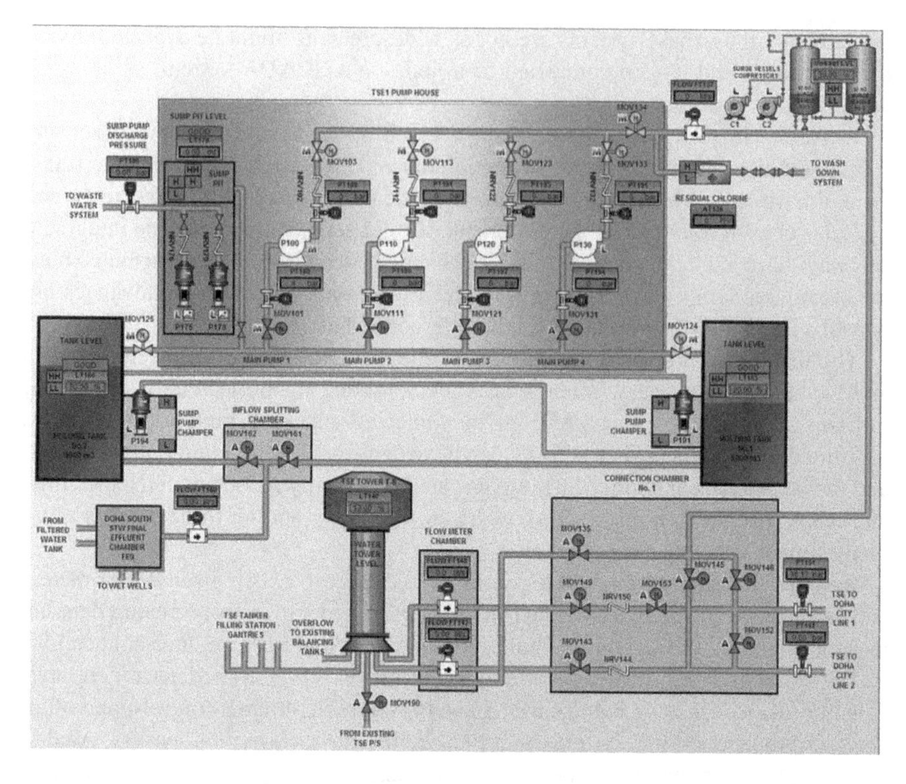

FIGURE 6.4 Example diagram of a supervisory control and data acquisition (SCADA) system used to connect and control various pumps, wet wells, flow meters, and other devices at a pump station. Today's IoT devices aim to replace expensive hardwired connections with easily deployable wireless technology. Public Domain: https://commons.wikimedia.org/wiki/File:SCADA_PUMPING_STATION_1.jpg

that does not support these programs, leading to efficiency gains that only scratch the tip of the iceberg. Trained data scientists can generate long-term storage plans, intelligent compression systems, and identify interesting conclusions or ideas that traditional thinking might obfuscate.

Further, there is value in using these systems as inputs into "Big Data" schemes. In contrast to "Small Data" – datasets that can be digested by an individual using manual tools – Big Data is a field of study that focuses on extremely large data sets that are beyond the scope of an individual's comprehension. These often constitute billions of records and require intensive computing systems for analysis – something enabled by the spread of elastic cloud technologies, where servers can be provisioned and scaled up on demand. Municipalities can pay for a few minutes on a supercomputer, rather than the infrastructure of physically constructing and supporting such a computer.

By logging data over time and retaining information that may not have an obvious utility, municipalities are setting themselves up to capitalize on conclusions drawn from Big Data platforms. For example, the Dutch water authority, Vitens, faced embarrassment when it took workers more than 3 hours to locate and repair a large leak. Media coverage dubbed the fiasco the "Leak of Leeuwarden". By installing a system of sensors and capturing data on the pressure, flow, conductivity, geographic distribution, and manufacturer information, Vitens was able to harness the power of Big Data. They turned this data set over to a consulting group – CGI – that used a data scientist to begin analyzing this dataset in order to dramatically reduce the time required to locate future leaks.

While most of us think about data analysis in terms of looking for conclusions that jump out, Big Data works in a very different way. Data scientists focus on structuring data in such a way that multiple sources can be easily linked together – formats are compatible and common linking variables such as time stamps are identified (Figure 6.5). Then, they structure a series of queries on a massively parallel computing system, which slices and dices this data. These systems use statistical techniques like genetic annealing, where the computer will look for trends regardless of physical labels, make a small change to weighting, and try again and again until it arrives at a set of conclusions.

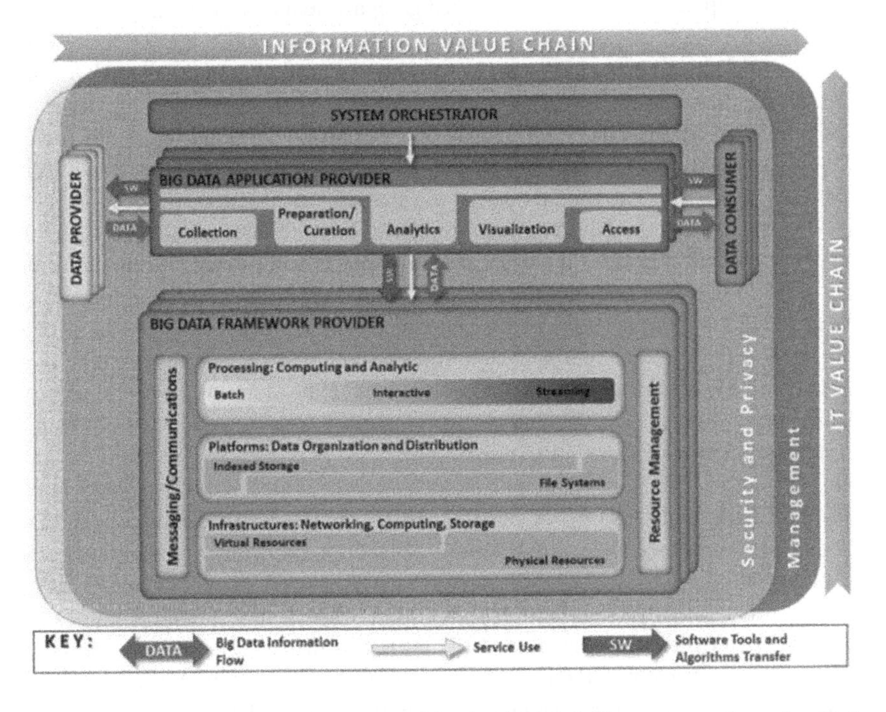

FIGURE 6.5 While Big Data is still a rapidly developing field, some questions about integrating information from multiple vendors are already being asked. The National Institute of Standards and Technology has proposed at least one reference architecture to promote interoperability – something that would benefit government agencies looking to avoid vendor lock-in. Public Domain: https://nvlpubs.nist.gov/nistpubs/SpecialPublications/NIST.SP.1500-2r2.pdf

These conclusions describe the historical data set extremely well and can include lists of temperature and pressure measurements just before line failure, manufacturer and installation data information, soil conductivity and acidity values if available, and more. This descriptive data set is not just an interesting historical footnote – instead, it forms the basis of a system of predictive analytics. When similar conditions are detected in the field again, alerts for preventative maintenance can be automatically generated and crews can prevent failures from ever occurring, in a form of artificial intelligence. Later, academics may study these models and determine the root cause of these failures – e.g. perhaps a nuance in the ductile failure mode of iron – but the system can identify when failures are going to occur, even if the engineering reason for this is not immediately clear.

While the idea of repairing problems before they become manifest may sound like the subject of science fiction, Big Data has established itself as a proven technology. Connecting large data sets with computing resources can enable companies to immediately generate value using today's technology. The previous case study may have made some skeptical, as the mention of a consulting firm can conjure certain budgetary expectations. Municipalities should consider partnering with local universities – computer science and engineering departments are constantly looking for ways to prove the utility of these computational models, and a large public sector data set could give them a chance to show off new techniques and give back to local communities simultaneously.

A final note on Big Data and university collaborations. Many utilities directors and municipal government figures loathe the public disclosure that comes with university collaboration. The thought is that the state of the system could be a source of embarrassment that could have political ramifications or lead to public outcry. While every brand of local politics is different, the reality is that university collaborations tend to bring positive attention to a water system – especially if that connection is actively sought out. Water utilities are typically seen by ratepayers as typical government agencies – ripe with inefficiency and waste, and a few decades behind on technology. University partnerships can quickly advance the state-of-the-art and lead to positive press – which in turn can generate outreach from agencies and companies also trying to generate innovative devices.

In Arlington, TX, the municipal departments of water and wastewater began working closely with faculty at the University of Texas. Advances in condition assessment programs using custom robotic inspection tools quickly generated positive press, as the department was able to show that it was saving money by not replacing lines unnecessarily. This led to further work with "smart" manhole covers that helped prevent overflows. The state of the buried infrastructure was not the subject of the story – rather it was the innovative approaches adopted by Arlington's Water Utilities that generated attention, and public works officials have received accolades for their work to improve the system. In the era of Flint, MI, hoping problems go undiscovered does not seem to be a sustainable strategy, and municipalities should consider collaborating where possible to push the current level of technology.

ACOUSTIC DETECTION TECHNOLOGIES

While smart meters and big data present a tremendous amount of opportunity, they also represent medium- or long-term equipment upgrades. The process of budgeting for and installing sensors throughout a system can take several years to complete, and even then, can sometimes only initially provide indirect evidence of leaks. Actual technologies to pinpoint leaks in real time are needed to address maintenance concerns as they arise. Acoustic technologies have become some of the leading systems in this regard and have seen widespread use in the United States and Europe (Figure 6.6).

If one were to visualize on the molecular level the flow of water stemming from a pinhole leak in a main pressurized to 80 psi, the result would be chaos. Water molecules would be violently bouncing off one another and moving in all directions. While the macroscopic flow would be outward, on the microscopic scale eddies and collisions in all directions would be prevalent. This chaotic system is not just metaphorically noisy – it is physically noisy, as well. Pressurized water leaks produce a characteristic sound that can be detected from a distance. The qualities of the sound vary with the size and scope of the leak as well as the material of the surrounding host pipe. Some of this sound can be attributed to actual motion of the water, but some of it stems from discharging water vibrating the surrounding pipe material. Because of this, sound can be carried significant distances along the pipe itself or through the surrounding material.

These sounds can be detected from a variety of locations – from the surface, from the exterior of the pipe itself, or even from within, as small devices can enter a line and make *in situ* measurements as they travel through the system. Each of these approaches has strengths and weaknesses and is able to generate different types of information about leaks.

Traditional acoustic sensors may be classified as analog or digital systems. Analog systems replicate and amplify the detected sound waves. They are often more inexpensive than digital systems and constitute the vast amount of older equipment in

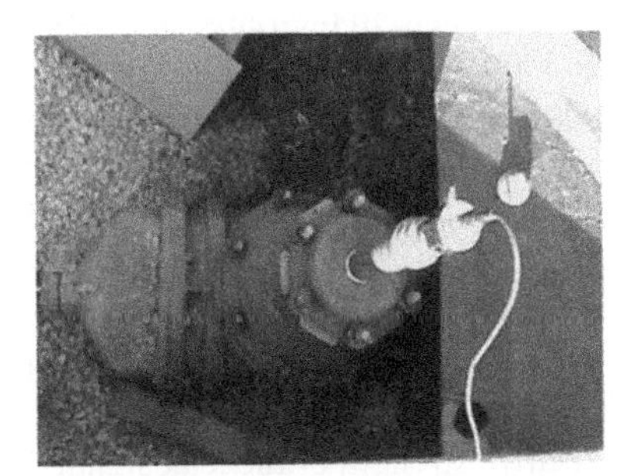

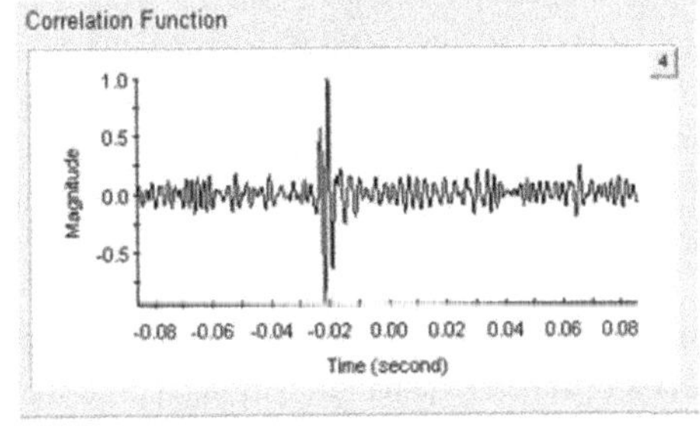

FIGURE 6.6 Installation of a hydrophone used in acoustic leak detection technologies (left). This larger unit was installed as part of a setup by Echologics, but smaller listening devices from other vendors operate under similar principles. Data (correlation function) indicating the likelihood of a leak can be seen in the figure at right – note the sharp peak. Public Domain: https://nepis.epa.gov/Exe/ZyPDF.cgi/P100EK7Q.PDF?Dockey=P100EK7Q.PDF

service. While they offer a near-infinite range of configuration and tuning options, they can be difficult to configure. Digital systems use a series of transistors, integrated circuits, or even software running on a microprocessor to filter or enhance a generated signal. Digital systems can be much more expensive than analog systems, but they offer a variety of presets that make them easy to use. For example, digital systems can be configured to quickly hone in on leaks in cast iron or PVC pipes with the touch of a button.

The simplest acoustic detection system is the listening stick. Listening sticks are typically passive devices that are composed of long metal rods with chambers or earpieces at the end. When the end of a listening stick comes in contact with a water line, it is able to transmit any acoustic noise from the line, through the metal part of the stick up to the interface. Sound waves are made up of vibrations, so these vibrations are easily transmitted through the uniform metal (usually stainless steel) of the stick, until they reach the earpiece where they are able to resonate as audio once again.

Listening sticks are extremely inexpensive and can be easily carried from location to location within a system. High-end units can be 4 or 5 feet in length, enabling them to easily reach water mains. Further, many do not have electrical components, meaning they can have service lives of years or decades, contingent on careful and considerate use. The downside to listening sticks is that while they may indicate the presence of the leak, they are not effective locating tools. Users may move along an easily accessible line to determine whether or not leak noises get louder or quieter but precision pinpointing is guesswork at best. Further, while high-end listening sticks may have electronic features that enable the use of custom headphones, all devices still require a considerable amount of end user training before a user can be productive. Finally, listening sticks require easy access to the water lines in question – deeply buried pipes or mains located under impermeable surfaces will be very difficult to listen to.

While expensive listening sticks promote built-in filtering capabilities, once trained, it is relatively easy for an operator to hone in on the sound of a leak. In PVC pipes, leaks typically generate a resonant sound somewhere in the range of 50–900 Hz, while leaks in metal pipes resonate between 500 and 1500 Hz – all of which is in the range of typical human hearing, even for older adults. While 50 Hz may seem low and difficult to detect, anyone who has ever heard the hum of an electrical current at night has successfully detected a sound in the 50–60 Hz range, depending on one's continental location.

While much can be gained from a listening stick, leak location requires a more sophisticated set of tools known as correlator systems. Correlator systems use a series of sensors distributed along a pipeline which feed audio signals into a processing device. The processing device analyzes the relative strength of the leak signal at each sensor, and determines the physical location of the leak so long as it is located within the sensor network. For example, ignoring flow characteristics, in a linear pipe, if a leak was located halfway between two acoustic sensors, its audio signal would have a similar amplitude in each. Flow characteristics and fluid dynamics mean that calculating the location of a leak involves much more than simple linear interpolation, but the example illustrates the point. An understanding of the distance between acoustic sensors is necessary for accurate leak location.

Acoustic sensors can take a variety of forms, and can resemble traditional audio equipment. Contact microphones are cheaper and easier to implement, but have a more limited range than the more expensive hydrophones. Vendors of leak detection systems will often advertise their maximum detection range – hydrophones, for example, can detect leaks up to 1,000 ft away. Use caution when interpreting these claims – location distances depend on both the magnitude of the leak and the pipe material, so this range may not be practical in application.

Advances in computing power have made it possible to combine the function of listening sticks and correlators in a single unit. Advanced listening sticks can acquire various samples of leak audio from points in a system. Each of these samples can be digitally tagged and stored on the device itself, referenced to a GPS location. The processing typically done by a real-time correlator can be performed in software on collected samples, allowing a leak to be located. Many of these devices even include special amplifying microphones for listening to leaks through soil or pavement for precise location between fixtures.

Ambient noise reduction can be a significant challenge. Fortunately, many devices include a variety of tunable filters to remove ambient noise from a signal. These filters can include high and low pass filters, which remove low and high frequencies, respectively; notch filters which remove specific frequencies – such as traffic or bird calls; or bandpass filters which can be tuned to allow only the leak noise to pass through for detection. This highlights a key advantage of digital systems: filters can be tweaked and altered in a variety of ways and remain constant over months or years of use. Older analog filters that depended on variable resistors and inductors could experience significant temperature effects and distortions with age.

Where exactly one listens to leak noise can have a significant impact on detection capability. As noted previously, some component of the noise is due to vibrations in the pipe wall itself, enabling effective transmission throughout the water system. This means contact microphones or listening sticks can touch almost any integrated part of the system in order to detect leaks. The water line itself and gate valves tend to produce excellent results, introducing minimal noise into the detection system. This is also true for hydrant gate valves. Some fixtures are material dependent. Shutoffs tend to be useful listening points in metal pipe, but amplify too much noise in PVC systems. Any location where the system leaves the buried environment and is exposed tends to be a source at which noise can be introduced – creek crossings are particularly notorious.

Sometimes it is difficult to access fixtures for effective listening. Fire hydrants can be convenient listening points, but introduce significant amounts of noise into the signal – especially dry barrel hydrants. Other devices can produce noise that resembles a leak. Pressure reducing valves and other size change fixtures produce turbulence that has a very similar quality and can be mistaken for – or drown out – the sound of a true leak. Further, in order to truly pinpoint the location of a leak, it is often necessary to make measurements between fixtures – hydrants, for example, are often 300–400′ apart. Ground microphones can be used in the space between fixtures, but they tend to have a very limited range and amplify the most ambient noise. Localize a leak as much as possible using fixtures and/or a correlator. Then use the ground microphone to hone in on the precise location for repair.

Remember that not all leaks can be detected. Some leaks are so small that they don't make any noise. Other leaks that have caused water to pool in the vicinity are inaudible as the surrounding water tends to muffle and absorb the sound of the leak. Some contractors will claim the capability to find *every* leak in a system – know that doing so is impossible, and hold municipal crews to a reasonable standard. However, if the contractor is willing to make payment contingent on finding each and every leak, it may be worthwhile to put their skills to the test.

The best way for municipalities to get better at finding leaks aligns well with best practices for reducing nonrevenue water consumption or losses. Leak detection should be a standard practice for evaluating the baseline condition of a system – not something that is only called upon when a ratepayer reports a drop in water pressure. Consider regularly testing all fixtures with acoustic equipment as part of a regular survey – for small systems, this could easily be part of an annual program. For larger systems, a multiyear effort may be more appropriate. These practices will give your crews more familiarity and expertise with the acoustic detection equipment, and help a system owner proactively identify leaks before they become full-fledged breaks.

When purchasing an acoustic leak detection system, think carefully about the materials in your system, spacing between access points, and the way in which you want to structure a leak detection program. For a program that focuses on emergency use, a high-quality listening stick with some embedded processing may be sufficient – something like the Aquaphon, by Sewerin. However, for more proactive leak detection programs, solutions that are capable of continuous monitoring may be desired. Echologics produces a suite of sensors and cloud processing equipment that can be retrofitted onto existing fire hydrants. This may be necessary if an area is experiencing frequent costly breaks or service disruptions.

ACOUSTIC INSPECTION TECHNOLOGIES

Anyone who has strained to hear the sound of a leak in a traditional acoustic detection system can attest that while it is certainly convenient to detect and locate leaks from the surface, doing so makes the process much more difficult. A variety of devices have been developed in order to harness the power of acoustic inspections in devices that can actually travel through a pipeline, detecting and locating leaks as they travel through the system.

The challenge in such an approach does not lie with leak detection, but rather localization. How can a device know when it detects a leak where exactly that leak is located? As was discussed in previous sections, even the best IMUs are subject to drift, making precise location a challenge. Some tools adopt a tethered approach. The Sahara system, developed by the Pressure Pipe Inspection Company, now a division of Xylem, uses this model. The Sahara consists of a small piezoelectric transducer mounted on a sensor head that is inserted into the pipeline through a tap. Once the device comes in contact with the flow, a parachute inflates to center and propel the device along with the flow. An above-ground operator controls the location of the probe by spooling out tether during the inspection. When acoustic anomalies are detected, the operator can move backward and forward until he feels the leak has been fully characterized (Figure 6.8).

The Sahara unit also contains LED lighting, an onboard camera, and a transmitter. The LED lighting and camera system enable the operator to see the context of potential leaks – is a crack visible? Is it part of a larger cluster of defects? Since the device is neutrally buoyant and pulled along by the parachute, it is not kicking up tuberculation or debris, making the video useable. The transmitter communicates with a man-portable locator at the surface so that significant leaks can be marked for digging and remediation as they are encountered. Further, the tethered nature of the system means that the precise location of the device in the pipeline is known at any time. The tether also acts as a securing device, preventing the probe from becoming lost or jammed in the system.

These units can be entered into lines that are actively in service. For a force main carrying sewage, this is not particularly significant, but on the water transmission and distribution side, this means that the system will come in contact with potable water. The probe and tether must be continuously disinfected to prevent the introduction of contaminating bacteria into the water supply.

The downside to tether-based systems is inspection speed and length. The Sahara typically comes with a tether that is slightly less than a mile in length, meaning that longer distance inspections will require multiple deployments. Further, in order for the operator to locate leaks at the surface, the device is typically engaged in relatively slow inspections, which can take a significant amount of time to complete.

Other devices eschew the tether, and can be freely and remotely deployed in pressurized systems in a manner similar to pigs. Once deployed, these devices can be retrieved at a downstream location and can inspect thousands of feet of pipeline, limited only by onboard battery life. Further, inspection speed is a function of flow rate, meaning that highly pressurized lines can be inspected very quickly. Two examples of autonomous devices that can perform acoustic inspections in pipelines are the Xylem (Formerly Pure Technologies) SmartBall and the MTA Pipe Inspector.

While both of these tools are touted as autonomous, it is important to note that they are not actively making decisions in the same way as a self-driving car or a robot exploring its surroundings – instead, they simply travel with the flow in a pressurized water main. Since these systems typically follow predictable paths, this usually does not present an issue – however, large changes in water consumption, such as suddenly open hydrants, can sometimes lead to these devices traveling in unintended directions.

These types of units are typically deployed through an existing tap or access point into the system, and travel downstream, collecting and aggregating acoustic information as they move through the system. The SmartBall tends to roll on the bottom of a service line, while the Pipe Inspector floats along with the flow. Both systems use similar acoustic sensors, but the Pipe Inspector has an added built-in CCTV camera, making it possible to obtain video inspection data, as well. Because the device travels with the flow, this video data is not always looking at areas of interest, but it can provide helpful context as leaks are encountered. At the conclusion of deployment, both units are retrieved via a net that is inserted into a downstream trap (Figure 6.7).

Because these *in situ* inspection units are traveling downstream with the flow, they are able to detect much more complex acoustic signals than just leaks – they can also detect the location of trapped air pockets. These air pockets typically occur at localized high points in the system, and can cause a variety of undesirable behaviors ranging from unpredictable flow, to energy loss when pumping, to air hammer effects or cavitations. In sewer systems, these air pockets present opportunities for hydrogen sulfide corrosion to occur. While venting systems can be installed to dissipate these air pockets, detection and location are necessary prerequisites.

How can these untethered systems accurately locate themselves in underground systems without GPS or other reference points? First, these systems tend to use multiple sets of high-quality accelerometers and gyroscopes, accounting in part for the high price of these units. Second, unlike driven crawlers, they tend to travel at relatively constant velocities, as they move with the flow. Further, the onboard accelerometers and gyroscopes can detect when major features like butterfly valves are passed through. Linear interpolation between the times at which these features are encountered can be used to determine an average velocity (Figure 6.9). With the Pipe Inspector, visual odometry can provide a check on the noted chainage, and both systems can be located via acoustic or electromagnetic technologies at the surface, if an external check on location is needed.

The downsides to these systems are primarily the cost, which limits their utility to large-diameter critical inspection mains. While pricing for these devices is typically offered as part of a professional services package, published responses to RFPs reveal that the cost to set up a SmartBall inspection can reach $25,000 per instance, and data collection fees of $15,000 per mile can be applied in addition. Surface location techniques can be much cheaper, especially if municipalities spread the cost of the detection equipment across its useful life.

Special care must be taken to not lose these tetherless inspection devices, and contract provisions should be included to spell out who is responsible for recovery and by what means. A SmartBall or Pipe Inspector lodged in some remote corner of a water distribution system can cause tremendous amounts of harm in a very short period. It goes without saying that any service provider should be both bonded and insured. While the location accuracy of these inspection techniques has been verified in a number of case studies, digging can be expensive, and contract provisions tying payment to accurate digs can be helpful. When excavating a repair, millimeter accuracy is hardly required, but locating a leak within a reasonable distance – say 10 feet – should be included as acceptance criteria for payment.

While this is not a text focusing on oil and gas inspection technologies, it has been interesting to observe some crossover into that industry. Free-flowing acoustic inspection devices tend to be much less expensive than the smart pigs that are used to inspect

FIGURE 6.7 The top graphic shows a conceptual illustration of the SmartBall by Pure Technologies. The bottom photos are of the insertion and extraction tubes used during an actual deployment – remember, it is critical to retrieve the device so that it doesn't travel to unexpected locations within the system. Public Domain: https://nepis.epa.gov/Exe/ZyPDF.cgi/P100EK7Q.PDF?Dockey=P100EK7Q.PDF

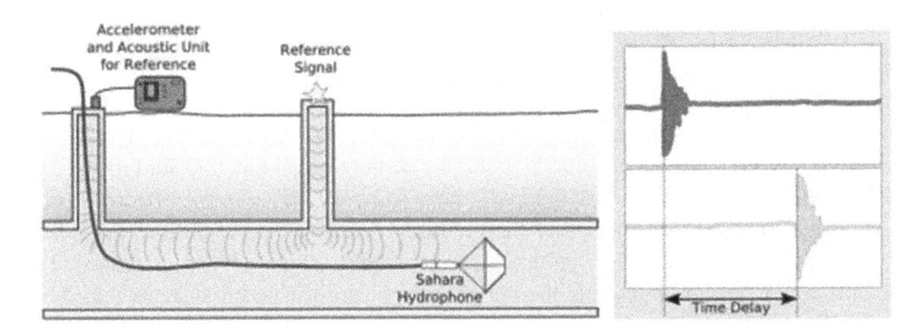

FIGURE 6.8 Illustration of the operation of the Sahara probe. Time between acoustic signal transmission and receipt is used to calculate wall thickness. Since the unit is tethered, it can be stopped to allow for further investigation of a specific area. Public Domain: https://nepis.epa.gov/Exe/ZyPDF.cgi/P100EK7Q.PDF?Dockey=P100EK7Q.PDF

Signal Generated by a Small Leak

Signal Generated by a Medium Leak

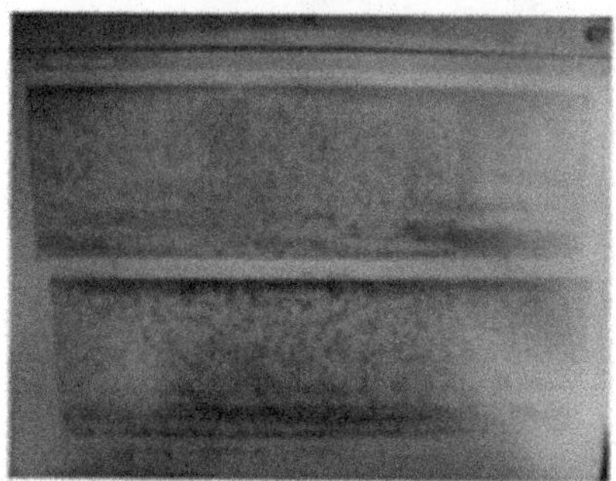
Signal Generated by a Large Leak

Signal Generated by the Discharge Line

FIGURE 6.9 Data generated by the Sahara device as shown on the operator console. Note the color coding makes it easy to identify areas for further investigation. Public Domain: https://nepis.epa.gov/Exe/ZyPDF.cgi/P100EK7Q.PDF?Dockey=P100EK7Q.PDF

large pipelines. While it may be tempting to leverage a water inspection tool in an oil or gas pipeline, these devices cannot be simply repurposed without modification. In oil pipelines, sensing units must travel longer distances, and survive and tolerate prolonged exposure to a variety of hydrocarbons. While it is obvious that devices must not malfunction when exposed to crude oil and other chemicals being transported, it is also critical to note that they must survive the buildup of thick layers of hydrocarbon and still be able to detect acoustic events. The packaging of the SmartBall for water inspections places the metallic device inside a protective foam exterior. This foam would almost certainly be unsuitable for deployment in an oil pipeline, and modification would be necessary.

It is also critical to think about flammability concerns when making the jump to oil and gas pipelines. One would be hard pressed to trigger an explosion inside a potable water system, and while theoretically explosions in sewer systems could be possible, they are extremely unlikely. In oil and gas pipelines, the possibility of an explosion is very real, especially in areas where air may become trapped or entrained into the system. Devices functioning in this environment must be rated to be either explosion proof or intrinsically safe, and achieving these certifications is not a matter of a few simple modifications. Intrinsically safe devices must be designed in such a way that all circuits in the device must be incapable of triggering thermal events that could ignite combustible material in air. Explosion proof devices must be contained within cases or housings that can withstand gas explosions, such that if gas were to somehow get into a material and trigger an explosion, the explosion would not be able to leave the device.

While this sets a high bar for devices to cross over when moving from the water industry into oil and gas, the reverse is not necessarily true. Perhaps devices that have been operating in oil and gas pipelines for years will begin to make their way to water and wastewater. While budgets of water systems tend to be much less than those of oil companies, there is still potential for technological crossover.

OTHER APPLICATIONS OF ACOUSTIC TECHNOLOGY

Pipelines constitute one-dimensional systems that can effectively convey acoustic signals over lengthy distances – a property that contributes significantly to the use of acoustic systems for leak detection. However, acoustic technologies also feature a number of uses beyond this that are worth mentioning briefly. Some of these techniques are useful for detecting blockages in gravity sewers, while others are useful for assessing the structural integrity of various types of pipe materials.

Gravity sewers typically operate under partially charged conditions, meaning that turbulence is a frequent occurrence in most pipelines. This means that listening for subtle signals such as leaks would be difficult over the noise of the nonlaminar flow. However, the headspace can serve as an effective conveyor of sound. If a sound wave is injected into the air above the flow, its clarity and amplitude will be impacted by how that space changes. An open space will transmit the signal fully and without distortion, while blockages or sags that impinge upon this space will lead to a reduction in the volume of the audio signal.

This observation suggests that audio signals could provide a simple way to testing whether or not sewer lines contain significant blockages. Sending an audio signal through the pipeline at one manhole and listening at another can quickly and immediately tell the operators if the line is free or obstructed. This premise forms the core of InfoSense's Sewer Line Rapid Assessment Tool, or SL-RAT. This transmissive acoustic inspection tool has recently had its technology formalized with ASTM specification F3220-17, and is seeing more and more adoption by municipalities (Figure 6.10).

When the SL-RAT detects that the received audio signal has been degraded in transmission, it uses an internal database of audio signals to determine the extent of the head loss present in the pipe. These signals were acquired from real-world use cases and include roots, cave ins, collapses, sags, grease deposits, and more. This process can occur almost instantaneously and results in an assessment that takes only seconds to complete.

It follows that transmissive acoustic techniques are faster than any visual sewer inspection. A transmitter is inserted into one manhole, the receiver is inserted into another, and in the time it takes for sound to travel from one to the other, the inspection is complete. Since sound waves lose intensity as a function of distance, the

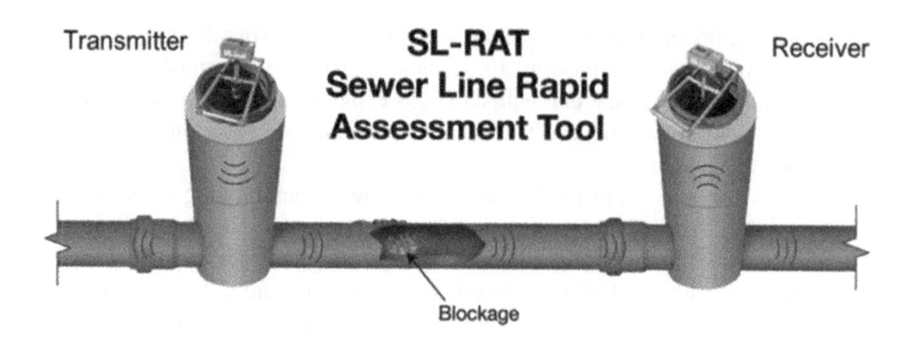

FIGURE 6.10 Conceptual graphic showing the operating principle behind the SL-RAT. Both the transmitter and receiver are installed in manholes. The signal sent by the transmitter is analyzed by the receiver in order to estimate blockages in the pipeline connecting the two units. Public Domain: https://nepis.epa.gov/Adobe/PDF/P100IY1P.pdf

distance between manholes must be known and entered into the device in order to accurately assess whether or not blockages have occurred. The SL-RAT accomplishes this by using onboard GPS chipsets to localize both the transmitter and receiver and 2.4 GHz radio signals to communicate between the units and synchronize internal clocks. Effectively, the system can leverage acoustics in order to inspect large portions of a sewer system with minimal traffic control (Figure 6.11).

The SL-RAT is not a replacement for traditional inspections, as it is not capable of accurately determining the reason for the lack of headspace – is it roots, debris, a sag in the line? Without visual confirmation, it is impossible to tell for certain. It does assign pipes a score ranging from 0 to 10. Scores of 0 represent complete blockage and indicate pipes in need of immediate attention. Scores of 10 represent lines that are completely free of blockages. These scores make transmissive acoustic techniques excellent triage tools for resource-constrained municipalities. If a comprehensive CCTV sewer inspection program must be spread over 5 years, in many cases this means that no information is collected on 4/5 of the system every year – meaning small problems can develop and grow to become major headaches by the time they are detected. Quickly assessing the remainder of the system using acoustics can determine which lines if any should be fully assessed prior to their location in a scheduled program.

Transmissive acoustic technologies can be faster and cheaper than zoom camera inspections – some estimates suggest that acoustic inspections can be completed in half the time as zoom camera inspections (3 minutes per setup) and cost less than $\frac{1}{7}$ the price when all costs are factored in (crew size, traffic control, equipment cost, etc.). That said, the technology is not suitable for large diameter lines and maxes out in pipes that are 18″ in diameter.

There are some caveats that must be noted. First, features other than blockages can cause the loss of acoustic signals – multiple capped laterals or offset joints can reduce the audio signal and give pipes lower scores. Further, while the tone generated by the SL-RAT travels down the pipeline, it can also travel through the air above ground. For long-distance measurements, this signal becomes sufficiently dissipated

FIGURE 6.11 Actual photos of the SL-RAT during transport (a) and deployment (b and c). Public Domain: https://nepis.epa.gov/Adobe/PDF/P100IY1P.pdf

so that it does not affect inspections. However, for shorter distances – less than 50 ft – this alternative pathway can have a significant impact on the received signal. The SL-RAT can detect this, and will indicate to crews that they need to move farther apart. Often this means moving to the next upstream structure and performing the inspection across an intermediate manhole.

While the SL-RAT is useful for gravity sewer systems constructed from a variety of materials, there is an additional acoustic technology that has been tailored to a specific type of pipe material. In 1942, a new type of pipe was constructed. This pipe used a concrete core as a structural element, surrounded by a steel cylinder and an array of prestressed wires that impart a uniform compressive force around the pipeline. These pre-stressed concrete pipes (PCCP) became extremely popular as pressurized water lines and have historically been very reliable. However, advances in metallurgy enabled manufacturers to produce high-strength, low-diameter wires, which were implemented in the 1960s and 1970s, without a thorough study of long-term longevity. Unfortunately, these wires have been failing prior to their design lifetime due to corrosion or hydrogen embrittlement.

When one wire snaps, stress is transferred to surrounding wires in the vicinity. If they have been deteriorating due to similar conditions, the pipe can rapidly fail. Unfortunately, these wires are embedded within the pipe itself – there is no easy way to visually inspect them and determine if they are at risk of failure. Previous inspections of PCCP pipe that were later known to have failed revealed pipe that was in generally good condition, with no obvious structural defects. Further, there is some evidence to suggest that even one wire breaking can be significant enough to trigger a massive rupture or line break through a cascading chain of failure.

These pipes have become ticking time bombs – and engineers have found that the best way to predict PCCP failure is to listen for the characteristic "ping" sound that is produced when one of these wires snaps – in many ways, the sound is similar to that of a guitar string breaking. However, detecting a single ping in a line that is in active service can be more difficult than finding a needle in a haystack. Not only must the line be monitored constantly, but the tolerance for error is minimal – each and every wire break must be detected in order for this to be an accurate failure prevention technique.

Historically, clunky arrays of hydrophones were utilized. Cables were run through each PCCP with hydrophones attached and spaced at 100–200′ intervals. These systems had a significant number of interconnected components, and the failure of any single hydrophone could render the entire system useless. Systems reading the data from hydrophone arrays were required to have limited spacing – each system needed to be spaced no more than 4–5 miles from the adjacent system.

Fortunately, advances in technology have obviated the need for hydrophone arrays. Recent advances in optical fiber have generated systems that rely on fiber optics – not just for transmission links between sensors, but also as the sensors themselves. As precise pulses of lasers are sent through fiber-optic lines, changes in the properties of the fiber optic cable can impact the transmission of light. These changes can be induced by physical deformation (strain gauges), temperature changes, or even acoustic signals. Acoustic signals actually exist as longitudinal waves propagating through a medium. In the case of sound, molecules are spaced closer and farther apart in a repeating pattern. In air, these pockets of compressed and dispersed molecules vibrate small bones in our ears that we recognize as tones or noises.

In a pipe, sound still functions the same way, except those pockets of molecules eventually strike the fiber optic cable, causing a very small amount of local deformation. This local deformation occurs in a very specific way that correlates with the frequency produced when wires break and causes some of the laser in the fiber optic cable to be backscattered to the source. The amount of time it takes for light to return to the source provides an extremely accurate distance measurement, and small laser pulses are able to detect sounds that last for less than a millisecond.

The result is that by stringing a fiber optic cable throughout the inside of a pipeline, a very detailed method of detecting and locating sounds is created. This technology is known as acoustic fiber optics, or AFO, and it has become the standard technique for detecting wire breaks in PCCP. AFO systems can detect sounds over a 100 dB range and can extend far beyond the reach of human hearing. Small mHz deviations as well as hundreds of kHz signals can be detected. Further, these wires can be installed in pipe that remains in service, using careful mounting techniques. While hydrophone arrays were limited by electrical attenuation, AFO systems rely on light signals that can travel for miles without degradation. Ranges of 12–24 miles for a single monitoring unit are not uncommon.

Unfortunately, these systems rely on a component of trust. While one can validate the claims of laser or SL-RAT data by performing a visual inspection, one must rely on the claims of manufacturers when deciding whether or not a system will be suitable. It is not possible to test an AFO system effectively outside of an actual failure event. Additionally, some thought must be given to false positives. Since a single broken wire can doom a section of pipeline, if a system errs on the side of caution and flags erroneous benign events, the result could be a series of costly shutdowns and replacement projects at significant cost. Further, AFO systems cannot detect other failure modes, such as corrosion of the steel cylinder or improper prestress in which gradual deterioration of the wires occurs.

That said, the Washington Suburban Sanitary Commission has extensively documented its use of AFO to observe the condition of more than 100 miles of PCCP. The system accurately detected more than 20 wire breaks before full failure, enabling the commission to quickly plan and execute a repair before emergency conditions developed. Each of these repairs cost between $250,000 and 500,000 depending on site preparation. A single emergency repair with property damage claims, bypass pumping, and other costs included would have exceeded $2.5 million. Thus, the commission claims that its AFO system has saved more than $42 million between 2010 and 2018. This is significant, as AFO systems are not inexpensive. The AFO system installed in Washington had an installation cost of approximately $21 million.

If PCCP can be such a ticking time bomb, and AFO systems are so expensive, why not just proactively replace all of these pipes before they fail? While many utility managers would sleep better at night if this were possible, replacement costs for a mile of water line can easily exceed $1,000,000 – especially in dense urban environments where prevailing wage requirements apply. Most municipalities try to chip away at this, but are not positioned for a massive capital project, when other older lines may be in need of imminent replacement. AFO monitoring systems seem to be striking the right balance at the present time.

7 Other Leak Detection Technologies

Focused Electrodes, Gas Tracers, and Infrared Imaging

While the previous chapter presented an introduction to the need for leak detection and an overview of acoustic technologies, there are many other systems available for detecting and quantifying leaks in pipelines. While these approaches do not have the market penetration of acoustic technologies, they can often accomplish similar results at a much lower cost. Further, some of these techniques specialize in sanitary sewers, where by detecting leaks of flow out of the host pipeline, they can also identify major sources of infiltration and inflow. Leak detection in these systems can be key to reducing and eliminating combined and sanitary sewer overflows (CSOs and SSOs, respectively).

SATELLITE SYSTEMS

With acoustic technologies, the accuracy of leak detection increases with the proximity of sensors to the leak itself. *In situ* technologies like the SmartBall can pinpoint leaks by rolling within a few inches of them, while correlators must make less specific estimates based on the relative strength of acoustic emissions. Yet, one of the most interesting advances in leak detection places as much distance as possible between the sensor and the actual leak itself. This extraterrestrial technology uses satellite imagery to detect leaks from space (Figure 7.1).

While this may seem counterintuitive, remote sensing techniques can provide valuable context for on-the-ground surveys. Drones equipped with lidar are able to see variations in topography that are masked by vegetation and detect ancient ruins buried by layers of jungle. Similar systems when mounted with microwave sensors can actually detect soil moisture levels and improve irrigation techniques for agriculture. Satellites can accomplish similar tasks over an even broader perspective. Whereas drone-based cameras can accurately survey a few acres on a single flight, satellite imagery can capture terrain or weather information for thousands of square miles at once (Figure 7.2).

While imagery can be useful, leak detection requires the use of radar to accurately detect water below the surface of the earth. "Radar" is a term that originally stood for radio detection and ranging (the same method of abbreviation formed the basis of the terms/acronyms for sonar and lidar). In typical radar systems, a radio wave is

FIGURE 7.1 A variety of satellites perform remote sensing of the earth's surface on a daily basis. Some produce standard imagery, while others are capable of detecting water through microwave radiation, such as instruments aboard NASA's Aqua satellite, shown above. Public Domain: https://science.nasa.gov/ems/06_microwaves

FIGURE 7.2 Some of the microwave images clearly show changes in moisture content as variations in intensity, while false coloration can help increase the visibility of relevant information. The left panel shows a C-band image of sea ice, the center panel is an L-band image of the Amazon river, and the right panel is an enhanced L-band image of the Rocky Mountains. Public Domain: https://science.nasa.gov/ems/06_microwaves

transmitted from an antenna, bounced off an object, and returned to a receiver. By timing pulses of radio emissions, the radar process can be used to perform time of flight measurements over long distances – and was used heavily by many countries during World War II.

After the war, scientists discovered that pulses sent with wavelengths of between 1 and 10 cm were particularly well suited to identifying precipitation. The degree of reflectivity of radio waves (not just the mere fact that they were reflected) could be used to determine the rate of precipitation as well as the area in which it was falling. Hence, weather radar became increasingly popular and can be seen by millions of Americans every day on local news broadcasts. Interestingly, many visualizations differentiate between different types of precipitation – rain, snow, sleet, etc. This is not usually detected by radar itself, but is added to the precipitation data in the post-processing stage, usually by adding information from other sources such as aviation weather stations or METAR reports.

Radar can use a broad spectrum of radio waves for analysis – ranging from 1 mm (approximately 300 GHz) to 10,000 km (approximately 30 Hz). Each of these waves is able to propagate through the atmosphere in different ways. Some are able to curve around the earth's surface; others can bounce off layers in the atmosphere; and still others are able to penetrate through clouds, buildings, and other structures. While the last descriptor may sound frightening, it is that property which gives us cellular phone reception inside our homes and businesses. For remote sensing applications, radar in the microwave band is typically used, with frequencies ranging from 300 GHz to 300 MHz and wavelengths from 1 mm to 1 m.

Microwaves are often classified by their band or range of frequencies. There are different standards for bands, often tied to national regulatory agencies, but the IEEE has developed a system of letter codes that are widely used in the industry today. S band frequencies range from 2 to 4 GHz, and include modern Wi-Fi, Bluetooth, and some forms of cellular communications. K band frequencies range from 18 to 26.5 GHz and are often used in automotive applications – adaptive cruise control radars and speed detection devices. Speeding drivers with older radar detectors may notice a high rate of false positives as newer vehicles with radar technology enter the roadways. X band frequencies at approximately 9 GHz can be used in spectroscopy or material identification applications.

For purposes of leak detection, microwaves in the L band of the spectrum are most commonly used. These waves have frequencies of 1–2 GHz and are also used by GPS satellites and amateur radio operators. They are notable in that they can penetrate clouds, vegetation, and even the surface of the earth to some extent, and L-band waves have been used by scientists to detect subsurface soil structures on Earth and water on distant planets. The depth of ground penetration varies as a function of frequency. Lower frequencies can penetrate deeper than higher frequencies. For example, a 1.0 GHz wave can penetrate wet sand to a depth of 4.9 m and dry sand to a depth of 9.0 m. A 1.5 GHz wave can penetrate the same materials to depths of 3.3 m and 6.0 m, respectively. Higher frequencies can carry more information (larger bandwidth) and respond more selectively to material properties – choosing the right frequency can often involve analyzing various tradeoffs for an optimal solution.

Radar reflections or backscatters are often impacted significantly by the dielectric constant of the material encountered. Dielectric constants are a measurement of how a material responds to an electric field. This property, which is also known as permittivity, describes the degree of polarization a material undergoes when it is exposed to electromagnetic waves. Materials with high permittivities or large dielectric constants are polarized much more than materials with lower dielectric constants. This principle is used regularly by electrical and electronics engineers, as selecting materials with high dielectric constants in electronic component design can be a key method of increasing capacitance.

The actual numerical dielectric constant is a unitless quantity that represents the ratio of the permittivity of a given material to the permittivity of a vacuum or free space. Though it is commonly encountered in dielectrics or insulators, all materials have a dielectric constant or relative permittivity. Just like density, this quantity is an intrinsic material property and can be used as a tool for material identification, as in spectroscopy applications. In water, the dielectric constant varies inversely with both temperature and salinity; that is, as temperature and salinity go up, the relative permittivity drops. At 20°C, the relative permittivity of water is 80.2.

This is significant, as L band microwaves are sensitive to a material's dielectric constant – one of the main reasons why they are widely used for identifying and mapping subsurface water flows. Dielectric constant roughly corresponds to the reflectivity of a material – materials such as vegetation or dry soil have low dielectric constants, typically ranging between 3 and 8, and do not significantly reflect L band radar. Water, with its dielectric constant of 80, strongly reflects radar waves and can appear as bright spots in a microwave image. While many materials also reflect radar (metal), subsurface water leaks tend to form unique spatial configurations that allow them to be readily differentiated.

Further, fine-tuned processing of variations in the dielectric constant can separate leaks from the water supply or sanitary sewers from standard ground water. Again, recall that the dielectric constant of a material changes as a function of salinity or the materials dissolved within it. Since fresh water, sewage, and naturally occurring groundwater all have different materials dissolved within them, it is possible to separate underground pools from all three sources using image and data processing techniques – some of which can be quite computationally expensive.

One additional note is in order when discussing satellite microwave radar technology. There is typically a tradeoff between aperture size or coverage and resolution. Radar systems that can map tens or hundreds of miles at a time must be captured with a large aperture. As a result, the resolution of an individual "pixel" tends to be quite broad – to the point where it may be of limited utility. Synthetic aperture radar (SAR) systems work around this problem by mounting a high resolution radar unit with a relatively small aperture on a moving platform. Waves are emitted as the unit moves along with the platform, and the returns can be stitched together as long as details of timing are known and recorded. This enables broad coverage of an area, while maintaining fine resolutions, so that information gathered can tie to a more precise location – a few feet rather than a few miles (Figure 7.3).

While the theory of these leak detection systems has existed for some time, the technology has only recently become commercially available. While NASA has developed satellite-based tools for detecting groundwater, and scientists can submit proposals to use these systems for research and development, usage remains highly technical. Some companies will also sell L-band SAR images – acquired by flight or satellite – to customers who wish to develop their own leak detection algorithms in the post-processing stage. MetaSensing is one of the companies that provides this service. However, doing so requires deep knowledge of radar, dielectric properties, and image processing and is beyond the capabilities of a typical civil engineer.

A more turnkey solution is offered by Utilis, an Israeli company that has developed its own algorithms and techniques for translating radar images into water-leaks. Utilis-IT has partnered with a number of traditional water service companies, including Hydromax in the United States and Suez in the United Kingdom to provide

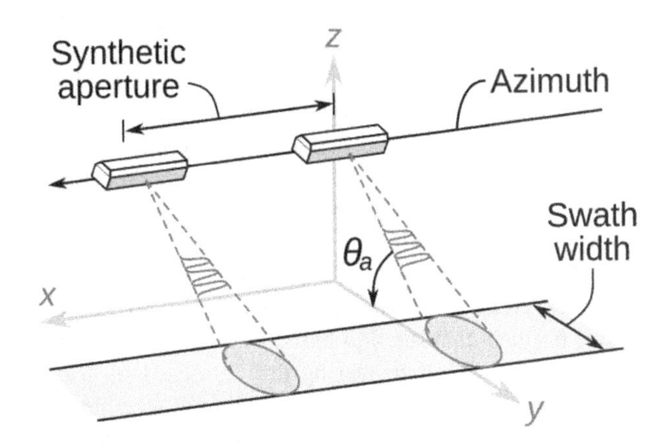

FIGURE 7.3 Diagram of the operational principles of a synthetic aperture radar system. Because these devices are installed on moving systems, the virtual antenna or aperture can be much longer than the physical antenna installed on the device. For effective usage, the spatial and temporal separation of signals must be precisely known. Image provided by RCraig09. Creative Commons: https://commons.wikimedia.org/wiki/File:Synthetic_Aperture_Radar.svg

satellite leak detection services as part of broader efforts to minimize nonrevenue water. While the business model of Utilis revolves around new technology, the company has been proactive in partnering with municipalities to publish case studies demonstrating the efficacy of the system.

This is due in large part to the fact that the technology can be easily verified. If the satellite system claims that there is a leak in an area, it is trivial to send a crew to confirm with traditional acoustic techniques and dig if necessary. Thus, it is impossible for the company to shy away from comparison testing. Further, while obtaining satellite imagery is not necessarily labor intensive, the cost of securing time on a satellite to scan a particular location can be quite high. Thus, it makes sense to spread the cost of a satellite analysis over a larger area if possible.

Some of the image processing and technical analysis the company performs on raw radar data involves performing several radiometric corrections – removing false responses that can appear when the radio signal bounces off buildings or other objects to produce stray results. The data is also overlaid atop traditional GIS mapping of the area and the transmission and distribution system components, in order to further refine probable results. With the Utilis system, each pixel corresponds to an area of 18 m^2, though advances in technology may reduce this to 3 m^2 in the near future. Leaks are localized in a radius, loosely corresponding to a confidence interval ranging from 1 to 100 m. This is sufficient to locate leaks at the level of the customer – a house or business.

Several comparisons have examined the efficacy of satellite leak detection systems against standard acoustic fixed installations. Interestingly, satellite systems locate far more "points of interest" than do acoustic systems. Each POI represents an area where leaks may be occurring. When standard crews are dispatched to investigate these POIs, more than twice as many leaks are found per point identified by satellite than by traditional acoustic methods. However, this does not mean that satellite detection services are a "slam dunk." Instead, they make the most sense for application in areas that do not have comprehensive leak detection programs, or in areas where the fixed cost of installing such a system would be relatively high. One California study found that Utilis made the most sense when the cost of a fixed-base acoustic detection system exceeded $68/mile on an annual basis.

It is difficult to give a one-size-fits-all recommendation for satellite-based leak detection services. While there have been a number of promising case studies, every service area has a different configuration of service lines, groundwater, and soil characteristics. It is unlikely that satellite-based leak detection services would work equally well in all areas. When considering this service, look for case studies in similar areas and speak to utility directors. How did the satellite detection project go? What were the shortcomings specific to that municipality? Do they plan to renew the service? How did they integrate it into their typical leak detection approaches? A little bit of investigation will quickly indicate if the technique is the right fit for a specific application. Utilis only began offering its service commercially in 2016, so it does not have the same volume of supporting data as acoustic techniques.

While Utilis is the dominant player in the space, other competitors are sprouting up. Satelytics, a company providing satellite-based image analysis tools to a variety of companies, has recently engaged in the water and wastewater space. Satelytics developed products for the oil and gas industry that combined a variety of satellite imagery in order to detect the leakage of hydrocarbon products as well as structural compromise to various transmission structures. It has developed a suite of tools for detecting contaminants in bodies of water and has recently expanded into water leak detection.

Of particular note is Satelytics's usage of artificial intelligence and machine learning to draw conclusions. While each type of satellite analysis typically requires the involvement of multiple experts to develop, Satelytics is reducing this time by relying on Big Data to assist. Not all correlations between different types of satellite imagery and physical phenomena are widely understood. However, by using large-scale computational resources to identify trends and by relying on expertise to refine those trends, Satelytics hopes to rapidly develop a suite of custom analytical tools for customers in a variety of industries.

Thus, while satellite leak detection is not ready to replace traditional crews, it is at a point where it is worth investigating, especially if favorable pilot pricing is offered as part of an investigation. Remember, these satellite companies are working to expand their install base, and good press can be mutually beneficial. While everyone deserves to be paid for hard work and effort, pricing should reflect the risk of trying this new technology and be tied to demonstrable results. While costs increase in terms of the number of images and square miles that are covered, payment should be tied to the efficient detection of verifiable leaks.

It is also notable that the strengths of satellite-based leak detection systems seem to complement the weaknesses of acoustic detection technology. Where large pools of water tend to muffle the characteristic leak noise that acoustic systems rely on, microwaves would be most responsive to these subterranean areas of saturation. Perhaps with time, a comprehensive leak detection program will use both technologies.

One final note: this text has been referring to this technology as "satellite-based" because that is currently one of the most cost-effective ways to use SAR systems to map large areas. However, as time passes and the cost and size of these radar units decrease, it may become possible to mount the units on small aircrafts or drones, making it possible to conduct microwave surveys on a smaller scale with higher resolution. Advances in battery technology, microelectronic systems, and radio frequency technology are all trending in a direction that could enable this in the future – the question then becomes, when will progress make this cost-effective?

GAS TRACERS

One may intuit that a natural way to detect leaks in water pipelines is to look for pooled areas of water – however, the area in which water tends to pool can be far removed from the area of a leak. Imagine a pipeline traveling up a steep hill. If this pipeline has a significant leak, water will flow out of the pipeline and continue to the bottom of the hill where it will likely collect at the bottom. If an investigation followed the flow of water, analysis would begin at the bottom of the hill, instead of the actual leak location. Gas tracer technology seeks to overcome this.

When a leak in a pipeline opens up, water tends to flow from the opening and continues flowing in a method directed by gravity, soil porosity, and other geotechnical factors. If a gas were to escape from this same location, the gas molecules would flow outward and immediately rise up at the point of the leak, assuming the average weight of the gas is lighter than air. This concept forms the basis of gas tracer systems. While in sewer systems, simple smoke can be used to visually identify the locations of broken pipes or improper connections, some consideration must be given to the actual gas deployed in waterlines (Figure 7.4).

The concept of using chemical sniffers arose with the advent of hydrocarbon transmission – as soon as mankind began moving flammable gas compounds around, it became critical to detect leaks quickly and efficiently. While flammable materials could be relatively safe within a pipeline, unexpected discharge could lead to buildup, explosions, and serious damage to property and life. One of the first gas detectors was the human nose. Some of the early commercially available gases were produced as a byproduct of the coal industry and contained sulfur compounds, giving them distinctive odors. Later processes for gas production made purer, odorless substances – which were extremely dangerous, as leaks were difficult to detect. In the 1880s, gas was artificially scented with sulfur in order to make leaks instantly detectable to humans in the area.

It is critical to note that while the sulfur smell is present in almost all natural gas that is used today, it is not an intrinsic property of the gas itself. Instead, gas companies are required to add odorization compounds that make the gas detectable

FIGURE 7.4 Smoke testing can be a quick and easy way to locate leaks in water and wastewater lines. However, to the public, the visual appearance of smoke coming out of the ground can be cause for alarm. Make sure to notify fire and police departments and local residents prior to testing. Creative Commons: https://commons.wikimedia.org/wiki/File:Smoke_Testing,_726_Emma_Ave,_Springdale,_AR.jpg

to average individuals at concentrations of $\frac{1}{5}$ of the lower explosive limit. These regulations came about in the 1930s and 1940s after a precipitating incident. In New London, Texas, the local school switched gas suppliers to save money, and used odorless waste gas for heating. Unfortunately, this gas built up in a crawlspace under the building and the resultant explosion completely destroyed the school and killed almost 300 people. Regulations mandating the odorization of gas quickly followed.

While the human nose is useful at detecting the presence of odorized gas, specially developed sensors – termed hydrocarbon sniffers – can not only detect the presence of gas (without added odorization), but also the concentrations present in a given area. These sniffers operate under electrochemical principles that are similar to those described in the section on hydrogen sulfide detection, and have become a quick and easy way to detect gas leaks. Interestingly, acoustic techniques can also be used to detect gas leaks, but require ultrasonic sensors due to the high frequencies produced when gas escapes through a leak.

The successful use of gas sensors in the hydrocarbon industry gave rise to the notion of using tracer gases to detect leaks in waterlines. However, since these lines do not normally contain detectable gases, new substances must be injected into the system. Gases can be injected through several different methods, but almost all involve connecting a commercially available tank into a waterline and pressurizing the line. Lines can either be empty or full, but if full, they must be pressurized with added gas to 3–5 psi over operating pressure to ensure that the tracer gas is sufficiently injected into the system.

While the type of gas can play a significant role in its efficacy in leak detection, the very nature of the gaseous state of matter imparts certain benefits. First, gas molecules expand to fill the space in the environment, without cohesion. This makes individual gas molecules able to diffuse through very small openings. Effectively, leak detection with tracer gases can identify much smaller issues than conventional techniques. Leaks can even be detected before liquid is able to flow through them. Second, this technique is considered to be nondestructive. Even though gas molecules are flowing through openings, they are not accelerating or exacerbating any damage that may be present.

Different types of gas may be used to accomplish this pressurization. One of the most effective is sulfur hexafluoride or SF_6. SF_6 is an electrically inert gas that is used as an insulating medium in many electronics applications. It is nontoxic, but heavier than air, and is easily detectable in small quantities. If SF_6 is used to detect leaks, small holes must be drilled above the pipeline to provide areas for the gas to collect – otherwise, because of its high density, it will not rise through soil to the surface to enable detection. An array of sulfur hexafluoride detectors are commercially available – ranging from electrochemical sensors, to infrared cameras, to specialized laser absorption system.

While SF_6 is a strong candidate for finding leaks in water systems, two properties limit its utility. First, it is an expensive gas. A typical cylinder with the capacity for 80 ft^3 of gas at 3,000 psi can cost almost $800 to refill with SF6. In contrast, refilling a cylinder with ultra-high purity nitrogen costs less than $100. If many lines were to be investigated, it can be worthwhile to find a cheaper alternative to SF_6; however, if a utility already has detectors and is looking for leaks in a localized area, it could still be a worthwhile gas to use (Figure 7.5).

The second drawback to sulfur hexafluoride concerns its impact on the environment. While SF_6 is nontoxic and relatively safe for humans, it is one of the worst greenhouse gases that has ever been studied. The Intergovernmental Panel on Climate Change found that SF_6 caused more than 23,900 times the warming of a similar amount of carbon dioxide. Once dispersed, it can remain in the atmosphere for more than 3,000 years. Thus, various locations are beginning to regulate its usage. In Europe, sulfur hexafluoride is banned as a tracer gas in most applications, and the California Air Resources Board requires users of SF_6 to disclose annual sales and consumption as part of broader reduction programs. Thus, while it may be acceptable to use SF_6 as a tracer if physical equipment is already in place, long-term plans should seek other alternatives.

Gases that are lighter than air are natural candidates, due to the fact that they can travel upward through soil and do not require the drilling of collection holes. Perhaps the most well-known lighter-than-air gas is helium, which has been studied extensively for this purpose. Helium gas consists of single atoms that are completely unbound. As a noble gas, it is inherently stable and unreactive. It does not pose a corrosion threat, and the small size of its atoms makes it able to identify some of the tiniest leaks. While we may think of helium as the substance filling balloons at children's birthday parties, it is also in demand for cooling large electromagnets, such as those used in Magnetic Resonance Imaging (MRI) machines.

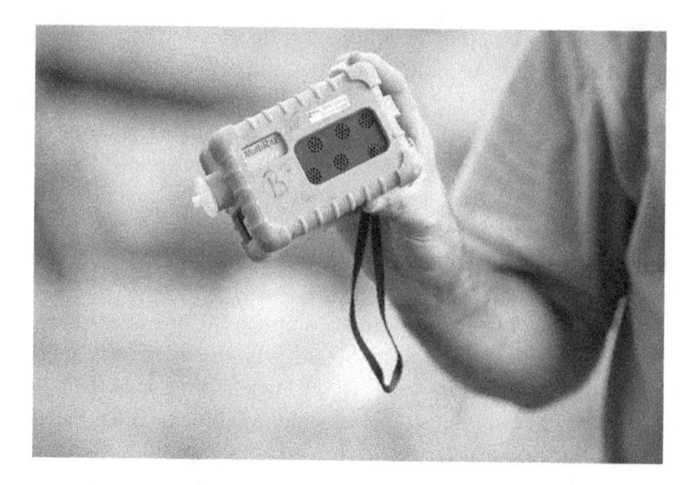

FIGURE 7.5 Portable gas detectors can easily and quickly detect a variety of compounds. Many examples exist for the hydrocarbon and hazardous materials industries, but other examples suitable for inert compounds are also available. Public Domain: https://commons.wikimedia.org/wiki/File:FEMA_-_38503_-_Hazardous_materials_gas_detector_in_Texas.jpg

Helium is mined from natural gas wells and is not regularly produced by industrial processes. When it is released, it rises in the atmosphere, never to be recaptured. Further, helium cannot be effectively stored for long periods in its gaseous form – it has a tendency to quickly leak out of small openings in even well-sealed containers. As a result, the market for helium can be volatile – spikes in consumption or reductions in supply can quickly lead to shortages. Further, when helium is used in leak detection applications, it has a tendency to quickly rise to the surface and then spread out, as it "sticks" to neighboring particles. It is not uncommon for gas molecules to wander between 10 and 50 ft from the site of the leak. This can make it very difficult to use helium to accurately pinpoint the precise location of leaks in buried pipelines.

Because of this, the de facto standard for tracer gases is hydrogen – the same material responsible for the explosion of the Hindenburg. Hydrogen gas molecules are smaller than helium, move faster than helium, and can just as easily escape through small holes (in fact, hydrogen molecules are so small that they can sometimes diffuse into the spaces between metallic atoms, weakening structures in a phenomenon known as hydrogen embrittlement). Significant concerns with hydrogen include its reactivity and flammability. Nobody wants to inject hydrogen into a water transmission system and cause an explosion.

This problem is addressed by dilluting pure hydrogen. A readily available mixture of 5% hydrogen and 95% nitrogen is used for this purpose. This mixture is classified as nonflammable and is very inexpensive. A size 80 hydrogen–nitrogen mixture can be refilled for approximately $50, making it very affordable. Further, electrochemical sniffers that can detect hydrogen are widely available.

While simple detectors can detect the presence of leaks near gas pipelines, more specialized quantification systems can be helpful to pinpoint the location of leaks.

A sensor that describes gas concentrations with a numerical value can make it easy to hone in on the exact location of a leak by looking for an area of maximum concentration. Many devices include probes with a variety of shapes and sizes, including wheeled sets for use on pavement, localized probes that can be pressed into the ground, and broader sensors intended for sweeping just above the surface. Many of these sensors can detect concentrations as low as 0.1 ppm.

Hydrogen does not exhibit the pooling that helium does, and it does not require drilling holes like sulfur hexafluoride. Thus, when hydrogen is detected in a large quantity, it is often very near to the actual leak location – within 3–6 ft, depending on the properties of the soil and other factors.

Detection devices also include small gas generators for calibration and bump testing. While hydrogen sulfide detectors often include small samples of the gas for this purpose, hydrogen can also be generated on demand. If the generator is filled with water and an electric current is passed through it, pure hydrogen gas can be collected at one electrode and oxygen at the other. Since most electrochemical sensors are solid state, they tend to have long lives, but do require frequent testing and calibration.

Hydrogen gas sensors can be purchased from manufacturers such as Sewerin, Inficon, Schoonover, and others. Equipment specifically designed for the water industry will often have several ease-of-use improvements – such as thresholds tuned to concentrations found in waterlines, and probes that are compatible with commonly encountered surfaces such as grass and pavement. The cost of these detectors can vary widely, but more expensive units tend to offer more features. Cheaper units may literally have a hollow tube containing the detector and a series of LEDs indicating a relative detection intensity. More expensive units can have LCD screens that display concentrations, memory modules that allow a user to save readings to the unit, and interfaces that can be linked to GPS systems for ease of localization – a useful feature when the individual locating the leak is not necessarily the same as the person performing the repair.

While useful, tracer gas systems tend to be more labor intensive than acoustic leak detection tools – they require some isolation of the line in question and pressurization with a gas before detection can even occur. With acoustic methods, no system preparation needs to occur before measurements can be acquired. Despite the preparation requirements, tracer gases remain popular as there are some scenarios where acoustic tools simply do not work. For example, pressure readings may indicate that there is a leak somewhere nearby, but a stream crossing or bridge can introduce so much noise that the leak cannot clearly be heard. Tracer gas systems could be employed in this setting to quickly find the leak and locate the area to be repaired. In other cases, if a line is out of service and not under water pressure, acoustic techniques cannot be applied, as there is no stream of water moving through the leak. In this case, tracer gas testing can be used to quickly pressurize the line and check for leaks, prior to reconnection to the system.

Thus, it is prudent for municipalities and water authorities to at least have a tracer gas system. It may not be used every day, and acoustic tools may still be the preferred method of detecting a leak, but when a leak cannot be located via noise alone, a tracer gas system is a much more convenient alternative to digging up a large section of line.

FOCUSED ELECTRODE LEAK LOCATION (FELL)

While much attention has been given to classifying the defects present in sewer lines, many municipalities are under intensive regulatory pressure to make their systems perform acceptably. A regular component of EPA consent decrees is the elimination of sanitary sewer overflows and a reduction in combined sewer overflows. While the reasons for combined sewer overflows are well understood – they are designed into the systems to some degree – sanitary sewer overflows should not occur in a well-designed system. After all, the amount of wastewater being produced should flow into properly sized interceptors, which discharge at a wastewater treatment plant that has the capacity to handle peak flows. Yet, surcharged pipes, loosened manhole covers, and smelly basements continue to occur, even in systems that are free of blockages.

The problem can often be attributed to leaks. Where leaks in water distribution systems tend to allow water to flow out of the system and into the surrounding area, leaks in a gravity sewer system allow surrounding water to flow into the system, overwhelming its capacity in a wet weather event. In other words, leaks or cracks can make a sanitary sewer system into one that becomes effectively combined through the presence of flaws that allow storm and groundwater to easily enter the flow and exceed the design capacity of the system.

While "infiltration" and "inflow" are often used in the same breath (and even have a combined acronym: I/I), there are significant differences between the two issues. Inflow refers to water that enters a sanitary system through intentional connections. These can include sump pumps from basements, gutters, or drains that are illegally connected into a sanitary sewer system. Some of these connections may have been established prior to effective regulation, while some are the recent work of unscrupulous property owners and contractors. Identifying sources of inflow can be quickly achieved through dye testing.

In dye testing, a tablet of fluorescent dye is placed in a suspect drain, and the drain is inundated with water. A CCTV system in the sewer observes to see whether or not the dye is washed into the main line. If no dye is discharged, the drain is legitimate. If dye comes out of the lateral, then the property owner can be fined for their illegal connection and forced to remediate (sometimes at significant cost). The CCTV crawler and dye test crew will slowly make their way down a street, checking connections from house to house.

Infiltration is much more insidious. Infiltration occurs when water enters a sewer system through defects – cracks, offset joints, broken tap connections, and other problems that may be identified in CCTV surveys. Unfortunately, unless surveys are done during wet weather events, the extent to which these defects contribute to infiltration can go unnoticed. For example, a CCTV inspection may notice an offset joint, but any infiltration that may come from that defect can be difficult to detect in dry weather. Some defects that contribute to infiltration might not even be noticed by CCTV surveys, such as cracks around intruding taps. While exfiltration can be a problem in its own right, contaminating local bodies of water, it tends to receive much less attention than infiltration, except in severe cases.

There are several schools of thought on how much infiltration is acceptable. Some argue that any infiltration should be eliminated – as even if the amount is not enough to cause SSOs, it still represents an increase in flow to the wastewater treatment plant and a waste in ratepayer funds, treating flow that should never have been in the system. Others claim that so long as overflows are not occurring, the system is fine, as some infiltration will be inevitable in large and complex collection networks. Complicating this analysis, infiltration can vary seasonally and regionally. The EPA has offered a standard of 1500 gallons per day per inch of diameter per mile of pipe (i.e. 1500 gpd/idm) as a threshold for acceptable infiltration.

Interestingly, when municipalities attempt to reduce I/I as part of their efforts to eliminate SSOs, they tend to focus on large-scale repair techniques, one of the most significant of which is lining. Lining is a trenchless technology that is often much cheaper and less disruptive than digging up a pipeline and replacing it. Some liners are inserted in solid sections, while others are formed from chemical processes and expanded to fill the host pipe (known as cured-in-place or CIPP liners). Since many of these liners comprise polymers, and most polymers tend to form watertight surfaces, the idea that a lined pipe could have significant leakage was thought to be laughable.

In practice, this is not always the case. Consider (as an analogy), the process of making a pizza crust. After the dough has been kneaded and worked, it is spread out to form the base of the pie. Any imperfections in spreading that dough can lead to very thin areas of crust – areas so thin that sauce can pass through, even if the crust appears to be intact. The same is true for many liners. If anomalies occur in the chemical reaction or curing process, or physical damage to the liner occurs, small pinhole leaks can form, which can be difficult to detect using traditional acoustic techniques, since sewer lines tend to be partially charged with "noisy" flow and because the liner is often completely encased in host pipe, making it difficult for characteristic acoustic vibrations to travel long distances for detection.

Municipalities request that contractors line their pipes to eliminate I/I and contractors visually verify the result of the rehabilitation using CCTV – a process that does not effectively verify whether the proposed rehabilitation has actually eliminated the source of the leaks! This can be compounded when the lining contractor and CCTV inspection contractor are the same entity and have an incentive to not find defects in the installed liner.

A new technology has arisen to investigate this issue as well as the issue of detecting I/I in sewer systems – focused electrode leak location or FELL (sometimes known by the trade name Electro Scan). The technology focuses on the electrical conductivity of wastewater streams (Figure 7.6). While pure water is technically an insulator, water with any dissolved ions in it quickly becomes a conductive substance – hence the reason we evacuate swimming pools during a thunderstorm. A current flowing through a wastewater stream will be present anywhere that stream happens to flow while energized – meaning if it passes through a leak, it should be possible to detect on the outside of the pipe.

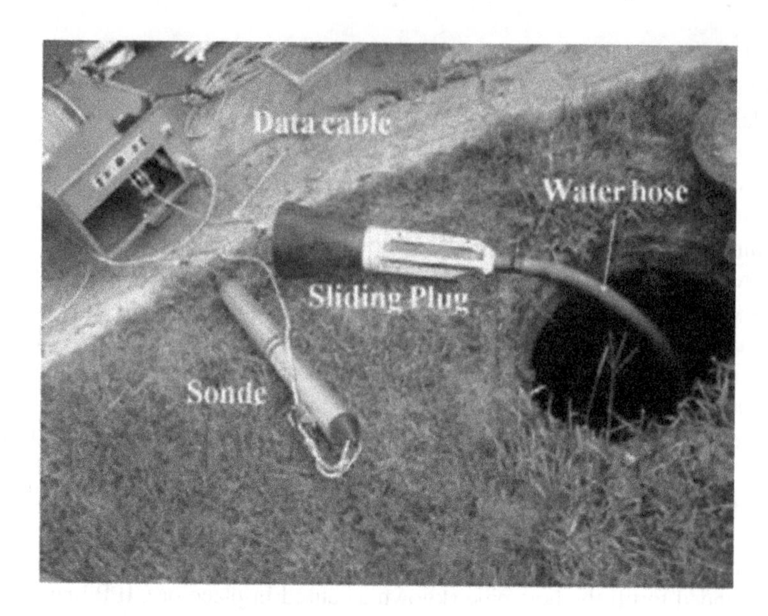

FIGURE 7.6 Major components of a FELL or Electro Scan inspection. Note the sliding plug connected to the hose that ensures the pipe in the vicinity of the sonde will remain charged to facilitate proper operation. Public Domain: https://nepis.epa.gov/Exe/ZyPDF.cgi/P100C8E0.PDF?Dockey=P100C8E0.PDF

The FELL system relies on this principle in operation. Unlike other technologies that can be self-contained in a single box truck, the FELL system must be deployed in conjunction with a jetter. To deploy, the jetter is sent from the downstream manhole to the upstream manhole and attached to the FELL probe via a specialized nozzle or cone. This cone creates a wall or column of water around the probe that remains in its immediate vicinity during deployment and covers the entire interior of the pipe wall from crown to invert. As the probe moves down the pipe, it is electrified. An external electrode is inserted into the surrounding soil to provide a return path.

Under normal operation, in an intact pipeline, there is no way for current to make its way from the electrode in the pipe to the return electrode in the soil without passing through the high-resistance pipe wall. As a result, minimal current makes it through the complete circuit. If a defect allows water to escape the confines of the pipe, electrical current will flow along with the water and complete a path to the return electrode. In this manner, leaks can be easily detected. As leaks increase in size, the amount of water able to pass through increases, as does the detected current. Thus, this system enables an operating crew to quickly detect and assess the magnitude of leaks.

While some electrodes have a camera system included in order to provide context of the surrounding pipe, this is not a typical stop-and-go CCTV operation. As the electrode is pulled through the pipe, changes in received current are automatically recorded along with the chainage of the device. Thus, inspections are completed quickly and detected leaks can be identified and measured without relying on judgment calls on the part of the operator or coder. Further, even invisible defects contributing to infiltration can be identified using this technique – imperfections below the flow (buried under silt) or not visible due to camera angles can all be captured using FELL.

This process can be used to evaluate whether or not lining projects were successful. A pre-lining FELL inspection can identify the location and magnitude of all leaks in the system. A post-lining inspection can be overlaid atop these results to show whether or not the defects causing leaks were successfully eliminated. Perfection is not always possible – certain taps may enable water to escape the system and appear as leaks under the temporarily surcharged conditions of a FELL inspection, but leaks occurring at cracks and joints should all be eliminated with a properly installed liner.

Since appropriate pipe lining involves both art and science, a FELL inspection can reveal whether or not visible imperfections are significant. For example, contractors may claim that some amount of blistering or wrinkling cannot be avoided – and that these superficial imperfections do not affect the barrier properties of a CIPP liner. Accelerant burns are another typical example, as these typically appear as spots of unusual coloration in CCTV inspections. They can either be harmless or provide a potential source of infiltration. A FELL inspection can quickly test these locations and determine which defects are contributing to leaks. As a result, many municipalities are including FELL tests as part of their post-lining acceptance testing.

FELL is only suitable for certain types of material – specifically nonconductive pipelines. Thus, while it is suitable for most types of plastic, reinforced concrete, and lined pipes, it cannot be used with ductile or cast iron pipes. The key principle of FELL relies on the pipe wall providing high resistance to an electric circuit and metal pipes make this impossible. It could be used to identify lining failures in conductive pipes, but this would not necessarily translate into leaks.

The published material for Electro Scan or FELL testing touts the system's ability to report the flow rate of fluid passing through each observed defect in gallons per minute. While these are useful 'back-of-the-envelope' calculations for estimating and quantifying infiltration, they should not be taken as real-world statistics tailored to a particular system. Instead, they estimate the amount of infiltration based on a 12″ head above the centerline of a pipe, and roughly classify defects as small, medium, or large. Small defects are estimated at allowing less than 1 gpm of infiltration, medium allow 1–4 gpm, and large defects allow more than 4 gpm. These assumptions are reasonable and Electro Scan has published several data sheets and case studies that clearly denote their usage, but wastewater professionals should know exactly where these statistics come from.

In many cases, FELL inspections are able to identify more defects than CCTV inspections, yet the former is not used as widely as the latter. Part of this is attributable to cost. FELL inspections typically require larger crew sizes to operate both the FELL probe and the electrode. Further, Electro Scan charges a fee for processing this data and delivering it to customers – approximately $2.00 per linear foot of inspection data. Make sure this fee is included in estimates or bids provided by contractors. Further, until recently, there was some level of discomfort with the purely

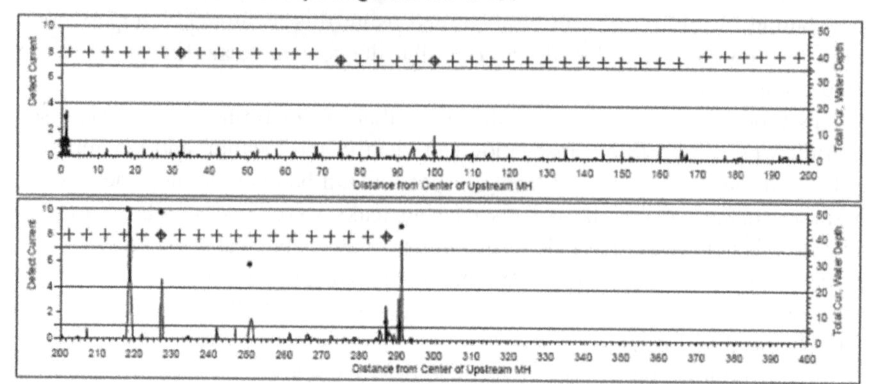

FIGURE 7.7 Sample data produced by the FELL system, showing distance traveled, water depth, and defect current, which can correlate to the magnitude of a water leak. Provided software can correlate these readings into loss or infiltration estimates. Larger peaks represent more significant leaks. Public Domain: https://nepis.epa.gov/Exe/ZyPDF.cgi/P100C8E0.PDF?Dockey=P100C8E0.PDF

quantitative nature of FELL inspections. Without "seeing" the defects the system observed, it was difficult for many wastewater engineers to truly understand the value of the data that was collected (Figure 7.7). Newer probes with integrated cameras alleviate this to some extent.

Further, while the actual electrical testing used by FELL is nondestructive, the overall process can cause damage to existing pipelines, especially if it is implemented as part of a pre-rehabilitation investigation. The surcharge caused by the cone of water surrounding the electrode places stress on the crown of the pipe and can lead to premature failure. While it is true that a pipe that fails during a FELL inspection was likely at imminent risk, collapses of deteriorated lines can and do occur and should not come as a total surprise.

A significant advantage of FELL inspections is that they can be performed easily in pipes that have not been cleaned. As long as the jet nozzle is able to make its way to the upstream manhole, the FELL probe can inspect pipes in a variety of conditions. Fats, deposits, and debris may obscure defects from a CCTV inspection, but any conductive path out of the pipeline will be captured by the FELL probe. This can represent a significant time and cost savings by eliminating expensive pre-inspection cleaning work. Further, while one may take issue with the blanket calculations used by FELL, the uniformity of the system does a good job of eliminating variation due to individual operators. Where different CCTV inspectors may disagree on what a defect should be classified as, or occasionally miss codes due to challenging field conditions, the FELL system eliminates these sources of operator error.

Many municipalities struggle to integrate FELL into an overall condition assessment program – it produces unique data that both competes with and complements CCTV inspections, and can take a chunk out of the overall budget due to its higher costs. That said, there are some clear cases where it should be utilized. First, if a municipality is specifically seeking to eliminate or reduce infiltration, Electro Scan can be money well spent. While CCTV inspections can identify a host of defects, FELL specifically identifies those that cause infiltration. Second, incorporating FELL into lining programs provides a standardized way to perform acceptance testing. For best results, both pre- and post-lining assessments should be conducted and compared to determine the efficacy of the lining operation.

FELL should not be used to completely replace CCTV inspections. While detecting defects that cause leaks can help to quickly reduce infiltration and make progress towards eliminating SSOs, traditional CCTV inspections provide other valuable condition assessment information – details about surface condition, spalling, root intrusions, grease deposits, and other conditions that may be observed by a FELL probe with integrated camera, but will not necessarily be focused on by the device. CCTV can provide a host of valuable details about structural and O&M conditions that the FELL probe is not designed to detect.

THE IMPORTANCE OF STRATEGIC PLANNING AND COST–BENEFIT ANALYSIS

Previous chapters have clearly shown that there are a variety of technologies available to quickly detect and locate leaks in both water and wastewater systems. Some of these systems are useful for emergency repair services, while others form the baseline for more comprehensive repair programs. The best programs will likely incorporate a variety of technologies from multiple vendors, as certain methods are able to detect leaks commonly missed by others. As a new generation of asset management professionals enters the water and wastewater field, it can be easy to fall prey to the temptations of data and spend money chasing sensors and information that can help develop more understanding of these complex systems. The EPA tends to provide guidance to wastewater collections system operators in the forms of policy, guidelines, and threats. Some general recommendations are offered here for those attempting to create their own cohesive leak detection program.

First, take some time to create a document outlining what you know about the state of your system today and share it with like-minded professionals. There is strength in numbers and you may find that your water authority is ahead of, behind, or on par with your peers. By beginning this dialogue, you may discover that your counterparts have solved particularly pesky problems, and can share their experiences with various contractors and technologies. You may find that there are certain actors to be avoided, certain vendors that work well with a strong verification regime, and certain suppliers who are truly reliable. You can also share pricing information, including any hidden fees or costs that came as a surprise.

Next, attempt to quantify any losses you are facing from the current state of your system. Some of these calculations will be relatively simple – such as estimates of nonrevenue water. Others may take some more legwork and estimation. How many emergency repairs do you perform due to line breaks? How much more do these emergency repairs cost than standard, planned projects? Has your agency had to pay claims for property damage due to events like sinkholes? This is important, because tying leak reduction programs to cost savings can ensure that technologies are a worthwhile investment. It doesn't make sense to pay for satellite detection or sensors on every fire hydrant unless the expense can be recouped through future savings. These calculations may surprise you.

For example, if a hypothetical municipality treats more than 12.5 million gallons of water every day, but only bills customers for 10.7 million gallons, at $1.50 per thousand gallons, these losses result in nearly $990,000 in "lost" revenue in a given year. Similar calculations can factor in the costs of emergency repairs and claims due to events like sinkholes. When everything is compiled, it may become clear that current leaks in the system cost the authority $3–5 million per year, a typical figure for a mid-size water distribution system.

Now, when beginning to roadmap target improvements, it is possible to estimate the threshold at which investments in technology begin to "pay for themselves". For example, in a system with $5 million in losses attributed to leaks, a $1 million program should reduce leaks by 20% in order to be a worthwhile investment. These savings do not necessarily need to take place on the same time scale – some asymmetry is expected as it takes time to train crews and implement a new technology. Small-scale pilot programs can be used to evaluate technology and mitigate risk. Many vendors will be willing to demonstrate a technology if they are in the area – take advantage of these services to see how certain techniques will perform under your real-world conditions. Don't become the utilities director defending excessive spending at a board meeting. New technology can gain both good press and bad press.

With that said, if your department is purely reactive, it is worthwhile to do whatever it takes to create a proactive leak detection program. While many of these programs find enough leaks to justify the investment in the first year, they also provide intangible benefits – they require crews to regularly contact all portions of a water distribution system and not just the problem areas. They also enable benchmarking and the ability to see changes over time. Staff will become more familiar with the entire system and begin to develop a sense of when things don't "seem right" in the entire service area.

This proactive approach could be implemented through a mixture of installed sensors, regular acoustic monitoring, and sporadic tracer gas inspections – the actual technological mix will depend on budget and specific needs. The important thing is to create and implement a program that assesses the entire system with regularity.

8 Condition Assessment in Water Lines

Interestingly, perhaps because of the involvement of the EPA in setting standards and practices for the sewer industry, much attention is focused on detecting issues in sewer lines before problems arise. While leak detection may seem to serve a similar purpose for pressurized waterlines, it begins by assessing a consequence of failure: that is, it only seeks to detect problems that are representative of lines that have already failed and are leaking. For some small-diameter distribution lines, this may be a reasonable approach. However, for large-diameter, high-pressure transmission mains, leaks can cause significant amounts of damage to the surrounding environment in a very short time and disrupt water service to ratepayers.

Thus, several techniques have been developed to assess the condition of pressurized waterlines before failure has occurred – these techniques look for hallmarks of deterioration and signs that could support preventative maintenance or proactive repair efforts. Some of these techniques look for actual damage to the pipe wall, while others examine more indirect indicators of failure. Many are deployed from within a pipe while it is in service, while others require dewatering or external access.

This section will give an overview of the various materials that are typically used in the construction of pressurized waterlines, deterioration mechanisms that can apply to each, as well as the condition assessment techniques that have been developed to detect signs or symptoms of degraded functionality. While there are many new and proprietary technologies, this section focuses on methods that have publicized their mechanism of action – black box "miracle" products are beyond the scope of this analysis, and any unproven technology should give water utility managers pause before adoption, unless it is part of a no-cost case study, pilot project, or limited demonstration.

MATERIALS AND FAILURE MODES

Even though water and wastewater lines can be constructed from the same materials, failure modes vary widely in application. Concrete pipes in wastewater applications must directly contend with hydrogen sulfide corrosion, whereas the absence of wastewater and the resultant bacterially active slime layers make H_2S attack much less common in concrete structures that are part of a potable water transmission and distribution system. Further, in many wastewater systems, corrosion and failure modes begin with an attack from the inside of the pipeline, whereas pressurized water pipes are frequently attacked externally from the surrounding soils.

Iron is one of the most dominant materials used in water supply systems around the world, and pipes with iron-based construction can have extraordinary longevity.

While it was not introduced in the United States until the 1810s, iron piping was installed in various locations in Europe well before then. A particularly famous cast iron pipe was installed before the French Revolution, during the reign of King Louis XIV. A major feature of the king's palace at Versailles was an array of water features and fountains – designed to dazzle members of the court, visitors, and foreign leaders. Unfortunately, the palace was not located near a significant enough water supply to meet these needs. An iron pipe was installed to carry water more than 15 miles to the palace and remained in service for more than 300 years.

It is not uncommon for iron pipe to have a service life of more than 100 years. The Ductile Iron Pipe Research Association (DIPRA) has even formed a "century club" to celebrate water utilities that have had lines in service for more than 100 years. While DIPRA is promoting the useful life of iron pipe, the extensive membership in this club – more than 578 utilities in the United States and Canada – calls attention to the age of water infrastructure. A further 29 utilities had iron pipe that was at least 150 years old as of January 2020.

In the United States, cast iron began to replace wooden pipes in water supply systems in the 1800s. Cast iron is named because it is cast in a mold. Early techniques cast sections of pipe in a vertical pit. While effective, depending on the length of the cast and environmental conditions, variations in material property could occur along the length of the section. The development of spin casting, where the mold is rotated while the molten iron is poured, did much to improve uniformity in pipe sections. Cast iron formed from these processes is sometimes termed "gray iron" due to the flakes of graphite embedded within the microstructure of the material (cast iron has a carbon content of 2.5–4%). These flakes offer no resistance to stress – a double-edged sword, as this property allows cast iron pipe to be easily machined, but also facilitates rapid brittle failure.

Cast iron was inexpensive, strong, and rated for the high pressures needed for water transmission and distribution systems. Unfortunately, transporting somewhat acidic water through these lines led to the onset of corrosion. When this occurs, discharged water has a red color, and flow capacity is reduced as tuberculation builds up on the inside of a pipeline, leading to a reduction in cross-sectional area. In the 1920s, a coating of cement mortar began to be installed on the inside of cast iron pipes. This coating effectively inhibited tuberculation, but led to significant leaching of calcium and hydroxide ions into the flow of water. Effective "seal coats" have been used to form a barrier between the transported flow and the cement mortar liners.

Thus, cast iron pipes with intact liners protected by seal coat are relatively safe from corrosion from within, though many lines currently in service do not have this full suite of protection. Corrosion can still attack pipes from the outside, especially if cast iron pipes are installed in acidic soil. Over time, this contact can lead to pitting and graphitization, which can result in leaks or failure of the line. Graphitization can be particularly insidious; these areas occur when the surrounding iron has been corroded away and a loose network of graphite and iron oxide remains. This network can sometimes maintain structural integrity for a short time before failing suddenly and unpredictably.

Ductile iron has almost completely replaced cast iron in the pipe manufacturing process. By adding small amounts of magnesium and carefully controlling casting conditions, carbon can be induced to form nodules of graphite, rather than sheets. These nodules give ductile iron pipe (DIP) tremendous strength advantages over cast iron – for a given thickness, some estimates suggest that DIP pipes are twice as strong as comparable cast iron pipes. Further, ductile iron is less likely to suddenly form a brittle fracture, as stress can no longer be conducted along graphite flakes. Since it became commercially available in the 1950s, DIP has completely replaced cast iron in new pipe construction.

That said, DIP tends to suffer from corrosion in similar ways to cast iron. To prevent corrosion from within, ductile iron pipes can also be installed with cement mortar liners. To prevent external corrosion, some DIP is installed with a protective external polyethylene wrapping that inhibits contact with acidic surroundings. While these wrappings can significantly inhibit external corrosion, they can also be damaged during installation or other underground work, leading to locations with sudden concentrations of corrosion and rapid failure (Figure 8.1).

Soil and water corrosion are not the only failure modes inherent to iron pressurized pipes. Manufacturing anomalies have been associated with multiple failures, especially in older cast iron mains, and installation issues can introduce small cracks and stress concentrators that can lead to premature failure. Fortunately, many failures in iron pipe begin with pinhole leaks that progress to larger failures. While this progression can sometimes be rapid, it at least gives the utility owner some time to identify and locate an issue prior to catastrophe.

While many forms of DIP are manufactured in a way that creates a protective oxide layer on both the inner and outer surfaces, cathodic protection systems can be

Note: arrow pointing to longitudinal propagation of crack

FIGURE 8.1 Photo of a catastrophic failure of cast iron pipe. Note the large defect and longitudinal propagation of the crack down the structure. This section of pipe was installed in the 1930s in Louisville, KY. Public Domain: https://nepis.epa.gov/Exe/ZyPDF.cgi/P100EK7Q.PDF?Dockey=P100EK7Q.PDF

installed to further slow the rate of corrosion. Outside of North America, bonded zinc coatings are commonly installed (and are part of design and installation standards) to further inhibit corrosion. Many of these same techniques can also be applied to steel pipe, which is much less common in water supply systems than DIP, and does not offer the same base level of corrosion resistance.

Another common pipe material used in pressurized lines is asbestos cement (AC), sometimes referred to by the trade name Transite. After 1900, large parts of the construction industry recognized that adding asbestos fibers to cement could have significant beneficial impacts on material properties. AC composites had improved tensile strength – sometimes on par with reinforced concrete – yet at much lighter weights, as asbestos fibers weighed a fraction of the amount of rebar cages. They also had improved resistance to corrosion, and AC was installed as roofing material, wallboard, and ductwork, in addition to piping (Figure 8.2). When the carcinogenic nature of asbestos began to be fully understood, these fibers were replaced with crystalline silica.

AC suffers from the material leaching noted in the discussion of cement mortar liners in iron pipes, and sometimes AC pipes were dipped into a bitumen sealant to protect the pipe material from internal and external corrosion, erosion, and abrasion. AC waterlines are particularly common in the American west, as cities and populations were expanding at the time this material was commonly available. The American Water Works Association estimates the useful life of AC waterlines as 65–105 years, depending on local conditions – meaning that many, if not all, in-service lines are nearing the end of their lifespans.

Now that asbestos is known to cause cancer in humans, many researchers have studied the impact of pipe degradation on the water supply. In addition to the

FIGURE 8.2 The use of asbestos cement was not limited to pipelines. Prefabricated panels were used in many forms of construction. This photograph shows an example of a post-war house with corrugated asbestos cement paneling. This particular structure is part of the Chiltern Open Air Museum. Image provided by Antony McCallum. Creative Commons: https://commons.wikimedia.org/wiki/File:Prefabfront.jpg

structural integrity issues that come with deteriorating pipes, AC lines provide an additional public health concern. Fortunately, the World Health Organization has performed ecological population studies and found that from an epidemiological perspective, asbestos leached from in-service pipelines does not produce an increase in the cancer risk to humans. While damaged pipes or newly installed pipes could release fibers in sufficient quantity to impact gastrointestinal systems, most danger from asbestos cement pipes is faced by the construction crews that must dig up, cut into, and remove these materials in the course of replacement.

While the deterioration of a cement mortar liner in a cast iron pipe may be troubling, it is not necessarily a harbinger of imminent failure. As the mortar liner leaches out, softens, and wears away, corrosion has an opportunity to begin in the host pipe – meaning there is still ample opportunity to detect and repair a problem. AC pipes do not have a similar margin of safety. Deterioration of the surrounding cement can pose a major structural hazard to the pipeline itself through rapid loss of structural integrity. Thus, many AC pipe failures are sudden, catastrophic, and comprehensive. When the cement material loses structural integrity, a large section of the pipe wall can suddenly collapse – there is no gradual progression from pinhole to failure.

Thus, the AC pipes currently installed in the water supply system can be seen as analogous to the prestressed concrete cylinder pipe (PCCP) ticking time bombs discussed in the acoustic inspection technologies section. These lines are all coming to the end of their service lives, failure can be catastrophic, and a proactive assessment program will be essential to prevent widespread failures from disrupting water supplies (Figures 8.3 and 8.4).

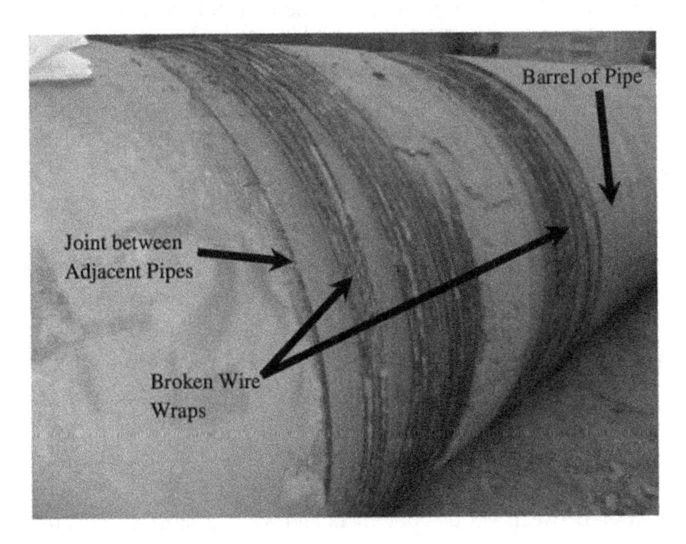

FIGURE 8.3 Photograph of a section of prestressed concrete cylinder pipe with arrows noting the location of broken wires. A few broken wires can quickly lead to catastrophic failure in these pipelines. They are often detected via acoustic measurement. Image taken from Hajali et al., *J Pipeline Syst Eng Pract.* STM (ASCE): https://ascelibrary.org/doi/pdf/10.1061/%28ASCE%29PS.1949-1204.0000219

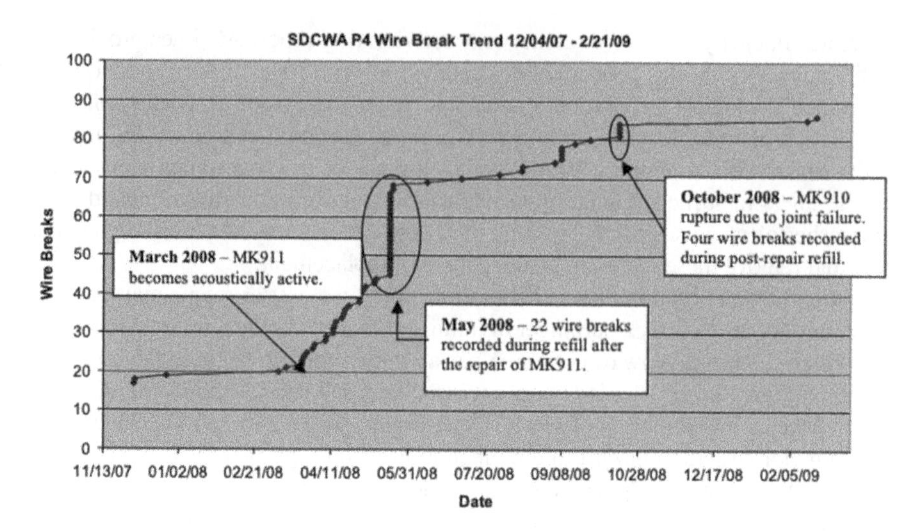

FIGURE 8.4 Graph showing the noise heard by acoustic fiber optic (AFO) system installed in a 72″ PCCP in San Diego. Note the pattern in May 2008 – breaks are often accompanied by other breaks, as structural load is distributed to adjacent wires, which may also be weakened. Image taken from Stroebele et al., *Pipelines.* STM (ASCE): https://ascelibrary.org/doi/pdf/10.1061/41138%28386%2968

Finally, more and more water mains are being constructed from polymer-based materials. Two of the most common are high-density polyethylene (HDPE) and polyvinyl chloride (PVC). While HDPE piping was produced as early as the 1940s, it did not become a standard for water supply systems in the United States until the late 1970s. Similarly, while PVC piping was invented in Germany in the 1930s, it did not become part of the American water supply network until the mid-1950s – due in large part to global events in the interim.

While polymer-based pipelines are generally extremely resistant to corrosion from both the surrounding environment and the fluids being transported, they have been known to fail prematurely, and engineers and materials scientists are continually working to develop more comprehensive understandings of the stresses and environmental conditions these pipes must withstand in daily, continuous use. For example, the production of polyethylene piping materials has advanced significantly since the origins of the technology. Early objects constructed from polyethylene were subject to UV damage and manufacturing defects that provided nucleation sites for crack growth. Other installations failed prematurely, as engineers failed to note that PE pipes can behave as viscoelastic materials and plastically deform when under constant stress.

The interaction between various polymers and water treatment systems is still being examined. A municipality in Ohio found that using chlorine dioxide in its water treatment system embrittled the HDPE lines throughout the system by producing a thin, oxidation layer throughout the system. This caused the installed lines to simultaneously begin failing more than 60 years before engineering estimates predicted.

Failures are not limited to HDPE – PVC pipes have also been shown to fail prematurely (Figure 8.5). One major differentiator between PVC and HDPE relates to

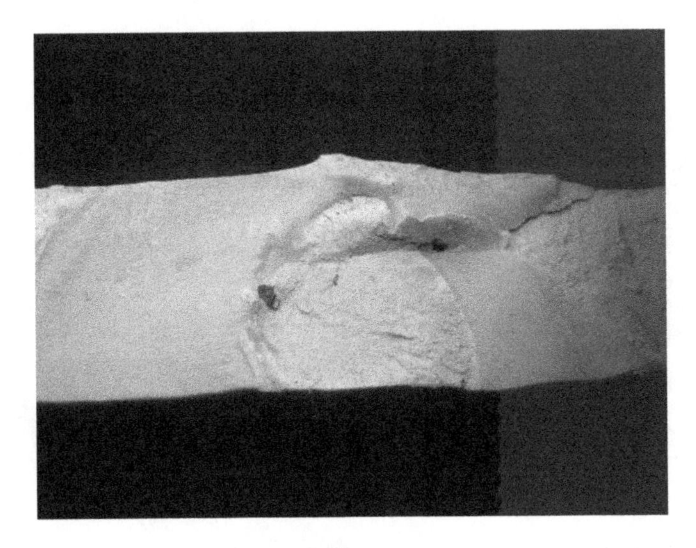

FIGURE 8.5 Poor quality PVC can suffer from inclusions of other minerals. These minerals can cause premature failure under pressurization. This photograph is from a 16″ transmission main that failed after 12 years of service. Image taken from Beamer et al., *Pipelines*. STM (ASCE): https://ascelibrary.org/doi/pdf/10.1061/41069%28360%2919

installation. PVC can be easily assembled via bell and spigot joints, whereas HDPE must be "welded" or butt-fused to produce long segments. While the welding process can introduce failures, it requires specialized equipment and knowledge, and is seldom attempted by novices. PVC, on the other hand, is more commonly assembled by individuals with insufficient training; installation errors can be a common source of failure.

An extremely common installation error is over-belling, where the spigot of one pipe is forced beyond the bell of its mate, causing the entire section to split and a large fracture to propagate longitudinally through the section. Other installation errors can stem from improper use of chemical cement, especially when segments are not held in place for the entire cure time. Excessive exposure to UV rays – by leaving a section of PVC pipe in the sun for extended periods – can lead to structural weakening and deficiencies that are not immediately apparent to the human eye. This means a careless installer could notice that some pipe intended for installation has been left in the sun for too long, cover up his mistake, and allow compromised material to be installed – with imperfections that would be invisible to even a trained inspector.

Some PVC can also suffer from manufacturing defects. Sections of pipe are produced by a powder extrusion methodology where precursor powder material is melted and extruded through a die under pressure – pasta makers (or Play Doh pasta makers) work in an analogous, albeit simpler, fashion. Sometimes plastic will not appropriately fuse back together after passing through the die – resulting in a pipe with trapped residual stress, that follows the site of improper fusion. These knit lines can cause pipes to spontaneously break or shatter longitudinally when dropped or otherwise shocked. Even if not subject to shock, these areas of stress concentration can lead to premature failure during service (Figure 8.6).

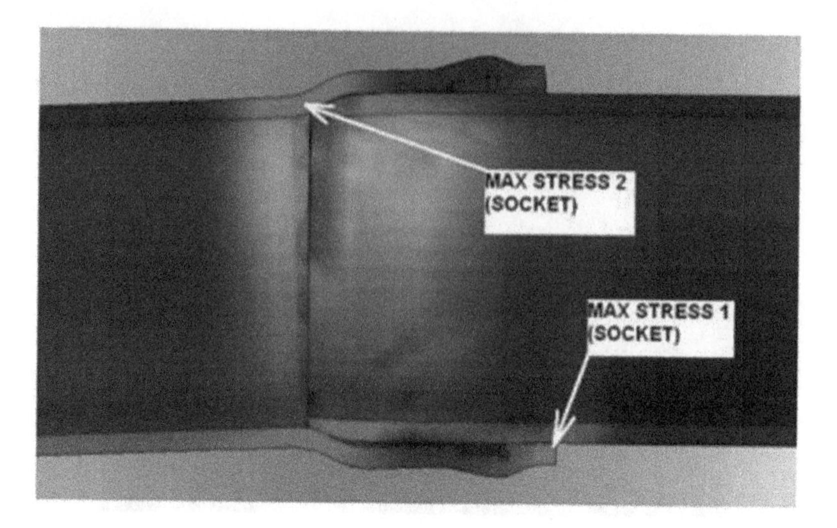

FIGURE 8.6 PVC pipe is extremely susceptible to failures stemming from installation issues. The above graphic shows the distribution of stresses in a PVC pipe during installation, with a small amount of deflection present. PVC is relatively inelastic, so stresses often result in cracking or other defects, rather than bending. Image taken from Youssef et al., *Pipelines.* STM (ASCE): https://ascelibrary.org/doi/pdf/10.1061/40994%28321%29110

Advances have been made in the manufacturing processes of both PVC and HDPE pipelines in recent years, and the industries are engaged in something of a rivalry, especially now that large diameter PVC pipes are available. While marketing materials emphasize relative advantages and disadvantages of each, unfortunately, municipal customers are still learning a lot about how these materials react in actual water supply conditions. Whenever possible, try installing a small section of pipeline to see how it will respond to your system's conditions, and be prepared to dig it up for a thorough investigation. If the previously-referenced municipality in Ohio had done something similar prior to installing 21 miles of HDPE, it might have had a chance to see the effects of chlorine dioxide in isolation and make a more informed material selection decision.

This mindset can be challenging for engineers to embrace – after all, municipalities are not science laboratories. Remember: materials engineers are not perfect – every advance in material properties that is well understood in the lab may have unintended consequences in the water supply system that are not visible for years. Working to understand new materials can reduce the risk and cost associated with new technologies, and make systems much safer, as a result.

VISUAL INSPECTION

It is clear that water supply systems could benefit from identifying and detecting evidence of failure prior to pipe leakage, and while many analyses jump to sophisticated technologies, visual inspection should not be dismissed. Quickly inserting a camera into a line to examine its condition can provide useful qualitative information about its condition and identify any evidence of corrosion.

For example, CCTV inspection of a water main may reveal the presence of tuberculation products. Contact with the camera system can indicate whether or not the mortar liner is beginning to soften, and missing sections of bitumen can be visibly identified with sufficient lighting. It is also possible to spot air pockets, cracks, or spalling, and other areas that should be investigated further. Some camera systems can be inserted into pressurized lines – either via a tether or parachute system. Full-scale crawler inspections are not common, as they tend to require dewatering of the pipe.

That said, a visual inspection does not provide the same value in a waterline that it does in a sewer line. Visible defects like cracks, offset joints, debris, and roots provide valuable insights into the condition and life expectancy of a sewer. The health of a waterline depends much more significantly on one factor: the remaining wall thickness and its structural integrity. From that perspective, a visual inspection can only provide clues or indirect evidence of this metric, and is not as comprehensive as it may be in the case of a sewer system. Still, it can be easily performed at a very low cost and may be worthwhile as an initial troubleshooting step.

DESTRUCTIVE TESTING

Perhaps the only true way to know how much life remains in a water transmission pipeline is to actually cut it out and examine the wall thickness. This process can also be performed in sewer lines, and involves removing a portion of the pipe wall known as a coupon and performing a more thorough analysis of its properties. As one would expect, replacing a coupon will require performing some kind of point repair to the host pipe – a process that can introduce additional points of potential failure.

Many vendors have developed coupon test systems that install representative metal strips in holders that can be subjected to the same conditions as the host pipe – sometimes these are threaded for easy removal and replacement, and are intended to minimize the risk associated with removing, testing, and analyzing coupons. Of course, slight changes in the environment of the coupon apparatus itself may make the samples it contains less than representative, so care must be taken when relying on these results.

Sometimes, systems may undergo hydrostatic testing, where a line is removed from service, isolated, filled with a test fluid, and pressurized to identify any fluid loss and corresponding defects. While in theory this process of pressure testing is nondestructive, it can cause pipe to fail at weak locations – identifying defects by causing them to fail prematurely. While it is more useful to find these defects during a pressure test than during routine service, sometimes these failures can turn a routine condition assessment into a major rehabilitation operation. Although this can be frustrating, lines that fail during such a pressure test were on the edge of failure in normal operations – in almost all cases where failure occurs during hydrostatic testing, a much more significant break would have been soon to follow.

NONDESTRUCTIVE TESTING METHODS

While destructive testing and visual inspection are useful, the former can be extremely expensive and time-consuming, and the latter can fail to yield enough actionable information. Further, since corrosion often begins on the outside of a water supply line, it is difficult to envision a method of direct observation that does not involve uncovering the line in order to check for pitting or other evidence of corrosion.

Many nondestructive technologies that can perform a more comprehensive inspection are material specific – that is, they can be used in ferrous pipes but not cement structures or vice-versa. Interestingly, many have not seen widespread adoption in the United States. This could be due in part to the fact that technologies only apply to a portion of waterlines, or the absence or regulatory pressures or other factors yet to be determined. However, just because nondestructive testing is not in widespread use does not mean that the status quo will remain acceptable.

Failures in pipelines tend to increase dramatically as structures reach the end of their service lives. A study in the midwestern United States over a 19-year period found that at the beginning of the survey, pipelines in the system under investigation experienced an average of 250 breaks on an annual basis. By the end of the study period, the number of breaks increased by almost a factor of 10, reaching 2,200 breaks per year. Each of these breaks was classified as significant and necessitated some kind of emergency response – including service disruptions and traffic control that caused economic disruptions beyond the direct cost of repairing the break. Thus, as buried infrastructure continues to age, regular condition assessment programs will only become more valuable.

In ferrous pipe – iron or steel – one of the most common forms of nondestructive testing involves remote field eddy currents (RFEC). Sometimes this is referred to as remote field technology, or RFT. The idea of eddy currents relies on the laws of electromagnetism, dating back to the experiments of Michael Faraday in the 1800s. In a notable experiment, Faraday discovered the link between electricity and magnetism. He found that when an electric current is passed through a coil of wire, a current can be produced in a nearby coil without any contact between the coils. Inducing this electric current led Faraday to name his discovery electromagnetic induction.

More specifically, when an alternating current is passed through a coil, it creates a phenomenon called flux – which is a measure of how the current is changing in time. This flux produces a magnetic field passing through the center of the coil. If another conductive structure is nearby, this magnetic field can create an electric current in the nearby structure – a phenomenon that was modeled by Maxwell in his famous equations. This principle gave rise to a number of devices that have changed the course of society including transformers, solenoids, electromagnets, and relays.

It is also a useful pipe inspection technology. In this application, an alternating current is passed through a coil that is termed the exciter. This exciter generates an electromagnetic field, as predicted by Faraday and Maxwell; however, this field also generates secondary fields in nearby conductive materials – notably the wall of a ferrous pipe. These fields produce circular currents known as eddy currents because of their similarity to eddies in the flow of water. The properties and dynamics of these eddy currents vary with the thickness and structure of the pipe wall.

While the field produced by the exciter coil generates eddy currents, it continues to travel through the pipe wall in a circular direction, eventually reentering at a separate location where it also generates additional eddy currents. A detector coil (usually ahead of the exciter) receives this signal. Changes in the phase and amplitude of the received signal are caused by the size and structure of the generated eddy currents, and can be interpreted to draw conclusions about pipe wall thickness. This is significant, as it provides a method of assessing both internal and external sources of corrosion through a measurement that does not require direct contact with any portion of the pipe surface – meaning buried lines can stay covered, and the electromagnetic field can effectively penetrate a cement mortar liner.

While the actual determinations of pipe wall thickness can involve complex calculations, the premise is that thinner walls will allow the electromagnetic field to reach the detector coil in a shorter amount of time with less attenuation than thicker walls would. This manifests itself in both a phase shift and a corresponding reduction in the overall amplitude of the signal.

Several commercially available devices that rely on this technology are available. Pica offers the Sea Snake and Hydra Sanke, which can be inserted into a line at a fire hydrant. These tools can navigate around 90° bends and be propelled by external pressure or water flow. They do require some preparation work – scraping off or swabbing excess tuberculation prior to an inspection, but can otherwise identify areas where the pipe wall may be compromised quickly and effectively.

Disadvantages of RFEC stem from both the technology itself and the deployment requirements. The swabbing of a line in preparation for an inspection can temporarily impact water color, leading to increased complaints for a utility. Further, the technology is not sensitive enough to detect extremely small pits (see ECT below), and when flaws are detected, they can be localized to their chainage, but RFEC does not provide information about circumferential or clock position. When repair typically involves digging, this may not be a significant limitation. RFEC devices must also be regularly calibrated – calibration records should be produced before the start of a project. Make sure they apply to the device being used, as variation can exist between individual units.

Conventional eddy current testing (ECT) can also be employed. In ECT, a single coil is used to inspect materials in order to look for specific, small-scale defects such as cracks or pitting. In this approach, the single coil still acts as an exciter; however, any significant change in the generated eddy currents will have an impact on the overall impedance of the coil. By scanning the coil across the surface, the location and type of various flaws can be detected (Figure 8.7). Arrays of coils can be used to increase productivity by allowing a larger area to be scanned simultaneously.

This approach is not widely used in waterlines, though it is commonly applied to piping – such as steam systems in nuclear power plants, or heat exchangers under extreme environmental conditions. ECT can also be used in petrochemical pipeline systems – both in transmission lines and at processing plants or refineries. It is typically much more labor intensive – and therefore expensive – than the RFEC systems used in water pipeline inspections.

A version of RFEC can be used in PCCP to detect broken wires. In this application, an electromagnetic field penetrates the concrete wall of the pipe and generates currents in both the steel cylinder and the prestressed wires. These wires can act as a transformer, and the effect has become known as RFEC with transformer coupling. By placing the detector coil at the opposite clock position in the pipe, the strength of the signal produced by the prestressed wire "coil" can be detected. A full complement of wires would produce a predictable electromagnetic field – but broken wires would produce a weaker signal – with the amount of attenuation increasing with the number of broken wires.

This requires careful mounting of the exciter and detector coils within the pipeline, and commonly available techniques benefit from manned entry to ensure appropriate positioning. This also requires dewatering which can increase inspection costs. Further, in order to accurately pick up on the transformer effect caused by the wires, both coils must be in very close proximity to the walls of the pipe.

Generally speaking, RFEC can provide accurate data on the condition of conductive pipe materials. It cannot be applied to AC or plastic pipes – and, perhaps more significantly, it cannot reveal any information about the condition of nonferrous liners or protective coatings. In a cast iron pipe with a cement mortar liner, RFEC can report on the condition of the host pipe, but cannot yield information about lining failure or delamination. While this may seem obvious, remember that many NDT inspections use specialized equipment – not a crawler, as would be the case in sewer inspections. Therefore, a camera is not along for the ride, and CCTV will not be generated as a matter of course. This means interpretation of these results is purely quantitative – there is no qualitative video to put defects in context.

A variant on this technology – termed broadband electromagnetic inspection (BEM) – seeks to provide a much more detailed map of the condition of a pipe

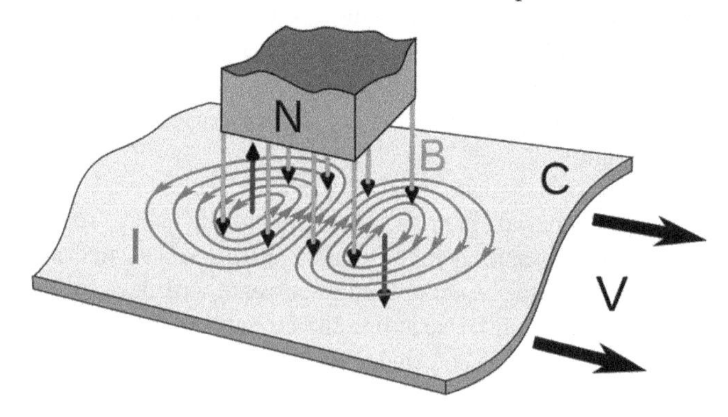

FIGURE 8.7 Illustration of eddy currents produced by a magnet placed near a conductive sheet. In this example, the conductive sheet is moving; however, in most pipeline inspection devices the mobile magnet is part of the inspection equipment and moves along the conductive pipe wall. Public Domain: https://commons.wikimedia.org/wiki/File:Eddy_currents_due_to_magnet.svg

wall. Rather than rely on a single emitter and detector coil to identify defects tied to chainage, BEM involves applying an array of electromagnetic sensors to a ferrous pipe. These sensors can be mounted on an external wrap or can be deployed on a device that moves through the pipeline. When the device – or droid – is inserted into the pipeline, it expands, bringing the sensor array into close proximity of the wall. The sensor is able to produce a circumferential map of anomalies in the pipe wall as it is deployed along the pipe. The droid typically moves down the pipe in 1 ft increments (similar to lidar stop scans), while the wrap must be removed and redeployed.

The result is a heat map – visually similar to those produced by laser/lidar inspections – that can plot the wall thickness of an iron or steel pipe throughout the entire structure, not just at a particular chainage. The leading producer of this technology – Rock Solid, an Australian firm, which has licensed their devices to others – refers to measured thicknesses as "apparent" wall thicknesses, unless they can be directly tied to a sample of pipe with known thickness against which the probe has been calibrated.

While this technique is useful, it can also be slow and expensive when applied to waterlines – especially for a fully internal inspection. It can have value when applied to form a statistical profile of a pipe. For example, if the condition of a critical water main is unknown, it could be exposed at several representative locations and wrapped with a BEM probe. The results of those inspections could be used to draw conclusions about the structural integrity of the line as a whole. While this may seem risky, points could be selected strategically – perhaps critical locations identified as part of a previous SmartBall inspection – as part of a follow-on analysis. As this detailed information can be obtained without the need for cleaning, bypass pumping, or any interruption in service, it can provide a high return on investment – for the right application.

While electromagnetic technologies may be useful in ferrous pipes, they cannot be applied to AC or cement structures, as eddy currents are not generated in insulating materials (no currents are – the very definition of an insulator). However, the principle of sending a signal through the pipe wall and measuring how its return changes as a function of thickness can be applied to other types of signals. Some ultrasonic sensors perform this task by sending high-frequency acoustic waves through a pipe wall. These waves will bounce off of a material change – such as the soil–pipe interface – and return to a receiver. The time between transmission and receipt can be used to measure wall thickness.

While an approach that does not rely on electromagnetic signals may seem well suited to asbestos cement, unfortunately water-logged or softened materials can seriously attenuate ultrasonic signals, making it much more difficult to determine remaining wall thickness. This softness can provide a useful measure of deterioration in its own sense, however. Tuberculation can also provide similar degrading effects through both scattering (due to the unique structure) and attenuation. As many utilities are reluctant to remove this buildup, this can seriously limit the utility of ultrasonic inspections. Further, insufficient bonds between materials can also trigger ultrasonic reflections, which can be misinterpreted as a lack of thickness. For example, if there is delamination between a liner and the host pipe, ultrasonic signals could be reflected off this irregular interface and erroneously a very thin section of wall.

Some recent advances have applied acoustic inspection techniques to determine the wall thickness in AC pipes. However, they rely on sending sound waves downstream through the pipe in order to make these measurements, in a manner that is not entirely dissimilar to the SL-RAT. This approach is based on the idea of acoustic propagation – suggesting that a sound wave traveling through a fluid behaves in part as something akin to a water hammer. It causes the fluid medium to expand and contract and apply stress to the surrounding pipe.

To apply these vibro-acoustic waves, a portion or the pipe or a fixture is struck by an object to produce a sound. Sensors are placed at known distances (typically 100–300 m) from the strike location, and the time it takes the sound to reach the sensors is recorded. As the pipe expands and contracts, the rate at which it transmits sound is proportional to its stiffness or Young's modulus. This means that nonstructural portions of the pipe – such as tuberculation deposits or cement that has been converted to gypsum – will not impact this measurement. Predetermined models map the time it takes for a sound wave to travel a fixed distance to various pipe wall thicknesses in order to determine the result.

While the technique sounds very academic, it has been applied in commercial products available from EchoLogics, and has been used in several pilot projects where the predictions of the acoustic model were tested against physical samples taken from pipes after inspection. These publicly available case studies from Las Vegas, Nevada, and Maple Ridge, British Columbia, validated the performance of the acoustic system.

In one interesting nuance, the acoustic system predicted that one section of AC pipeline had a thickness of 0.74″ – a result that surprised the investigators, as AC pipe rated for that size and pressure should have had a design specification of 0.66″, and there was no practical way that years in service could somehow increase the structural wall thickness, rather than decrease it. However, a manual dye test and expert analysis validated the findings to 0.01″ – showing the danger in relying on interior measurements and as-builts as proxies for thickness.

Other techniques hold promise, but have yet to be commercially scaled. Impact testing is one such method. In impact testing, a hammer is used to tap an AC surface, and the timbre and frequency can be analyzed to indicate structural integrity. High-pitched sounds are indicative of good structural integrity, while lower, dull sounds can indicate compromised structures. Arrays of transducers and automated hammers can formalize this to some degree, but it remains a manually intensive process. It is also used in other industries to evaluate ceramic oxide coatings inside pipelines, such as refractory coatings in catalytic crackers. Ground-penetrating radar can also be used to evaluate the thickness of AC pipes, but requires extremely careful application and expert interpretation.

One final technique worth mentioning also relies on electromagnetic signals and has been a stable of the hydrocarbon industry for years. Magnetic flux leakage (MFL) can only be applied to magnetic pipelines. In this technique, a section of the pipeline is magnetized by an external source, and a detector is placed somewhere in this region. Any flaws or imperfections will cause the magnetic field lines to change direction and "leak" from the structure. By measuring the magnitude of the magnetic field that leaves the material, flaws and defects can be located and measured.

This technique can be applied statically by moving sensors and magnets along the outside of a pipeline. However, it can also be installed on objects that move within the flow of the structure. In the oil and gas industry, very sophisticated pigs can be equipped with multiple MFL sensors that can evaluate the structural integrity of a pipeline as pressure forces the pig to move throughout the system – for miles and miles on end. Changes in the detected magnetic field correspond to more significant defects, so a larger signal corresponds to a more serious problem.

While this technique has seen widespread use in the oil and gas industry, it has yet to see widespread adoption in the water sector. Unfortunately, the market dynamics of oil and gas have created a situation where highly specialized MFL pigs are created and deployed as part of comprehensive pipeline integrity management programs. The customization, specialization, and cost of systems integration have relegated these tools to applications and industries with vast inspection budgets. The consequence of failure of an oil pipeline is more than an inconvenience – one event can rapidly cause tens or hundreds of millions of dollars in damage and fines. Waterline failures do not cause similarly scaled impacts. Perhaps with time, technology will become more commoditized as the cost decreases.

In the realm of plastic pipes, no nondestructive techniques have gained widespread acceptance. This is unfortunate, as it has restricted the applicability of plastic pipes in certain applications where regular testing and monitoring is a requirement – such as in nuclear facilities. As late as 2019, the Plastics Pipe Institute released a report stating that they took no position on the applicability of nondestructive testing to plastic pipe, but went so far as to discourage the application of existing techniques without modification.

To some extent, this view is warranted. Plastic pipes fail in ways that are very different from those of cast iron or asbestos cement. Corrosion is almost never a factor and wall loss of these polymers tends to be minimal. Failures occur at joints – due to improper assembly or stress concentration – or through crack propagation in the pipe material itself. Unfortunately, cracks tend to propagate very quickly in both HDPE and PVC, meaning that unless microscopic nucleation sites could be identified, there would be little value in looking for structural anomalies. Most of what could be detected would be significant enough to already constitute a problem, though visual inspection of HDPE welds could detect problem joints prior to failure.

Some devices exist that use ultrasonic arrays to evaluate the integrity of welds in plastic pipe, while other companies market inspection solutions that rely on different frequencies of electromagnetic radiation in order to accurately evaluate polymer-based pipelines. Microwave radiation (similar to that used in satellite leak detection systems) presents one route. In microwave systems, a transducer sends out a pulse of electromagnetic energy, and that energy is tuned so that changes in dielectric constant result in reflection back to the receiver. In this way, cracks, voids, inclusions, delaminations, and other defects could be detected as the dielectric constant changes as it moves from the host plastic material to the surrounding void. Evisive is one company that provides scanning solutions for plastic pipe using microwaves. Its devices have been used in a variety of plastic pipeline structures, though its technology seems more suited to high-tech applications than commodity waterlines.

STATISTICAL ANALYSES

Comprehensive inspection programs for pressure pipes can be difficult to structure. Unlike gravity sewers, there is no "one-size-fits-all" device for water supply systems. Whereas brick, reinforced concrete, and ductile iron gravity sewers can all benefit from a CCTV crawler, there is no single technology that can effectively provide useful condition assessment information for a water supply network consisting of a mixture of cast iron, ductile iron, asbestos cement, and high-density polyethylene pipes. Further, it can be difficult to draw conclusions across the industry, as lifespans of various pipelines are closely tied to installation and usage conditions. Some cast iron transmission mains remain in service more than 150 years after their initial installation, while flawed PVC distribution lines can fail a few years after being put in service.

One solution is to use software to try and identify trends in failures throughout a system. Unexpected water main breaks will always be a part of operating a water supply system, but analyzing these failures can yield useful, predictive information about lines that are likely to fail in the future. Effective analysis programs require logging every piece of known information about a pipeline – the more sources of data, the better. Fields could include the pipe material, vendor, lot, contractor, installation date, soil conditions, leak detection results, surrounding terrain, pressure/flow sensor data, repair history, and more. Often, when recent breaks are analyzed, common themes can be identified.

For example, although recent breaks may be scattered around a city, it may become clear that all breaks stem from a particular lot of plastic pipe, delivered on a certain date, or that lines installed by a specific contractor tend to fail prematurely. This analysis does not quite rise to the level of Big Data – a competent engineer provided with enough documentation could likely discover many of these conclusions with a homemade Excel sheet. Hiring a software developer who can integrate tools like the Python Data Analysis Library can also help automate the identification of correlations between variables.

Some software packages exist to help draw some of these conclusions – Fracta, Synergi, and other providers of reliability analysis software can help operators accurately determine the likelihood of failure of various pipelines – based on local trends and broader industry-wide data sources. These packages can also help calculate consequences of failure so that risk assessment programs can move beyond hypothetical assumptions to actual quantification of risk in terms of projected losses. These efforts can result in programs that may seem counterintuitive, but are actually very finely targeted – for example, inspecting recently installed PVC pipes more frequently than 80+ year-old cast iron mains. Such a program defies instinct, but could make sense if statistics show that the likelihood of failure of those PVC lines is highest in the two-year period immediately following installation.

Some tools offer insights that go beyond statistical correlations. Fracta has several algorithms that apply machine learning (ML) and artificial intelligence (AI) to predict lines that are going to fail based on models developed from ingested data. While AI/ML sound like terms that should exist within the realm of science fiction, they simply represent a series of algorithms that can improve automatically as experience is gained. In the case of pipe failures, these systems learn more about which pipes are going to fail in the future by looking at relationships between failures in the past. ML systems look at all possible correlations – even those humans might deem illogical – in order to develop extremely accurate predictive approaches.

Vendors of software platforms have an advantage over any single single water authority, as machine learning algorithms tend to become more and more accurate when they are trained on larger data sets. While a clever developer can implement machine learning on a set of citywide data, a cloud provider – managing data for dozens or hundreds of municipalities – can train their artificial intelligence systems on a dramatically increased scale, drawing conclusions about suppliers, materials, and even geographic regions that go beyond the capability of any individual or firm. The question of who "owns" condition assessment and failure data has not yet been resolved. If a municipality uploads their failure rates to one of these cloud platforms, can the cloud platform use the data to develop a model that it sells on the open market? The answer may be buried in an end-user license agreement that gets dismissed with an impatient click, but as AI becomes more advanced and data increases in value, much more attention will likely be paid to answering this question. Large municipalities could even monetize their repair data by selling this information to predictive analytics companies, offsetting the burden shouldered by ratepayers.

THE FUTURE OF WATERLINE INSPECTION

Historically, condition assessment programs have been of secondary importance for waterlines, compared to gravity sewers. Many municipalities engage in robust leak detection programs, but do not attempt to gauge the state of existing infrastructure prior to failure. A recent study by Utah State University indicated that only 45% of water utilities used some kind of condition assessment program for transmission and distribution lines, and those that did typically limited any techniques to large diameter lines in the transmission system. At the same time, that same study found that water main breaks increased by 27% over the six years studied.

As lines get closer and closer to the ends of their design lives, this problem will only worsen, unless massive amounts of capital spending are allocated to replace this aging infrastructure – and indeed, the American Water Works Association projects that more than $1.7 trillion will be needed by 2050 in order to replace outdated service lines and expand systems in areas of growing populations. To avoid service disruptions and reduce the costs of emergency repairs, expect to see many water authorities and municipalities turn to regular, periodic condition assessment of waterlines.

While historically these programs would have been financed by municipal bonds and rate increases, privatization of water systems is beginning to become more and more popular. Companies like American Water, Suez, and Aqua America are paying large fees to municipalities with the promise of more efficient service provision. While many of the long-term consequences of water and wastewater system privatization have yet to be determined, many of these companies are turning to comprehensive condition assessment programs, even before adopting a new system – reasoning that this information is critical to determine an appropriate offering price. As these companies seek to maximize the useful life of buried assets, condition assessment programs will become more and more popular, as many inspection technologies can be deployed for a fraction of the cost of replacement pipelines – a fact that will help return more value to shareholders.

9 Manholes, Laterals, and Ancillary Structures

Many collection system inspection projects place a significant amount of emphasis on the buried pipelines themselves – this has a certain amount of logic to it, as pipelines are some of the most numerous components in the system and a single failure can cause tremendous disruption. But just as problems with a road network can be caused by issues with structures like intersections, onramps, and parking lots, so too can ancillary structures impact the performance and overall health of a buried water or wastewater system.

There are many different types of structures – ranging from passive and simplistic junction boxes or weirs, to the complex and dynamic, such as manually or electronically actuated valves – each of which can have different condition assessment requirements. This section focuses on two of the most broadly applicable structures: manholes, which are key access points for many different types of buried utility systems, and laterals, which feed into sanitary sewer systems. Manholes and laterals are significant because technology and standards have emerged for comprehensively inspecting both, suggesting a future direction for the industry. Some practices applicable to other structures like buried chambers will be discussed where appropriate.

MANHOLES

Manholes have become ubiquitous parts of many buried utilities. Intended to provide convenient location for manned entry to support operational and maintenance activities, these structures have become synonymous with the public view of sewer systems, as they are often the only components that are readily visible from the surface (Figure 9.1). Manholes can also be associated with a variety of other buried utilities – ranging from fiber optics and telecommunications, to power distribution, to steam piping for industrial applications.

Many individuals spend a significant time working with a single municipal system within a particular geographic area, and have adopted the belief that there is a "standard" manhole – perhaps envisioning a precast concrete structure, or brick chamber, with a cover that is a fixed size. Yet, manholes vary significantly from region to region, depending on historical application and intended purpose. It is important to understand some of this context for a comprehensive understanding of condition assessment programs and technologies.

While it may seem obvious to provide points of manned entry for buried infrastructure systems, some early designers were so confident in their approaches that they believed this maintenance would be unnecessary for a generation or more.

FIGURE 9.1 Early manholes were explicitly designed around the dimensions of a man – enabling him to perform maintenance and inspection work. This diagram is taken from an annual report of a Massachusetts state agency in 1880. Public Domain: https://commons.wikimedia.org/wiki/File:Annual_report_of_the_State_Board_of_Health,_Lunacy,_and_Charity_of_Massachusetts_(1880)_(14781735722).jpg

Some of the first conveyance systems used gutters along side streets or separate ditches to promote the flow of waste – structures that came to be associated with unpleasant odors. These open systems attracted a variety of maintenance issues on a regular basis – tree branches introduced debris, people threw trash in the ditches, and vandals intentionally caused havoc by obstructing the flow.

At first glance, separate sanitary sewers appeared to be free of these issues. Designers intended for these systems to be "closed" – taking in waste from toilets and water closets, conveying it below the surface of the streets, and discharging (or later treating it) at a remote location. Debris could not be introduced without great effort, and man-sized access points would only introduce the opportunity for illicit access to the system. As larger access points would significantly increase the amount of trouble a determined individual could introduce into the system, early sewers were equipped with small lampholes, rather than larger portals.

These lampholes were structures 8–13″ in diameter, spaced at regular distances along sewer lines. As the name suggests, they were intended for inspection via lamp to view and assess flow characteristics. A worker could open the lamphole and lower a light down into the sewer to see whether or not the waste stream was flowing properly. Other members of the crew could open nearby lampholes and check for the presence of light – the absence of which could help pinpoint a blockage or obstruction. Flushing the system would help dislodge the debris and restore normal operation. Some lampholes contained small holes to allow for airflow, while others were sealed completely in order to prevent the introduction of trash and debris into the system.

In hindsight the problems with lampholes are obvious. Despite their confidence, early sewer designers had a lack of real-world data and made up for this lack with assumptions. Anyone with exposure to modern sewers knows the full extent of material that people will flush into the system. Victorian engineers simply couldn't imagine that the general public would view the sewers as a portal to get rid of embarrassing or unpleasant objects that would not flow in the ways system designers had predicted. Thus, when blockages inevitably occurred, the need for structures that could enable manned entry became apparent.

While lampholes have been rightly relegated to history, they persist in many parts of modern systems – though they are sometimes disguised within newer structures. While many lampholes were converted to manholes (at significant expense), others remained as they were constructed. As streets were repaved in the decades following system construction, some lampholes were buried, while others were preserved. In many cases, those that remained were topped with commodity manhole risers and covers, so that when these normal-looking structures are opened, a much smaller access point is revealed – sometimes causing a great deal of disappointment for the person making the discovery. That said, a lamphole is not useless – since it is larger than a common cleanout, jetters and vacuums can be deployed through a lamphole albeit with much more limited situational awareness.

When manholes began to be constructed, different locations took different approaches. In Europe, sewer systems were public works triumphs, and a great deal of emphasis was placed on not disrupting traffic for maintenance or repair activities. Separate chambers beside sewers were constructed – sometimes with elaborate spiral stairways – and connected to the pipeline itself via a walkway or passage near flow level. While these structures may not have inconvenienced the public in terms of travel, they placed an indirect strain on the general population in the form of costs, as constructing separate chambers was costly and building enough structures to support regular maintenance could quickly strain budgets.

Operationally, these systems were less than desirable. While modern readers may appreciate the lack of traffic control required with side chambers, the connections between the chambers and the sewers became prime locations for debris buildup, requiring frequent cleaning. Further, spiral staircases were convenient for manned entry, but were ill suited to the introduction of machinery. Hoses could not be easily deployed over the multiple twists and bends associated with these staircases, and large robots simply could not traverse the structure in the same way as humans.

Thus, manholes built off to the side of sewer systems have become historical curiosities. Even in Europe, where side chambers remained popular, manholes constructed directly atop or immediately adjacent to the sewer have become the de facto standard. Placing manholes directly atop wastewater lines (sometimes with a slight offset) allows manned crews and equipment to easily access pipelines. Further, cost can be reduced, as precast manholes can be constructed to accommodate common pipe sizes and geometries – and these systems can be installed in the same location as the pipelines themselves, cutting back on construction expense.

While some manholes are literally street-level openings into chambers or pipelines, most modern manholes have a standardized set of components to facilitate access. Using the same nomenclature to identify these components can dramatically improve the ease, standardization, and quality of condition assessment. These components describe the entire manhole "stack" – ranging from the incoming and outgoing pipelines at the bottom, to the cover at the top. Not all manholes will have the full set of components, and some – such as those permitting access to telecommunications equipment – will omit many of the features that are specific to pipelines (Figure 9.2).

At the top of all manholes lies a cover. Covers come in a variety of shapes, sizes, and materials. Common construction materials include cast iron, concrete, and fiberglass (or GRP: glass-reinforced polymer). Covers are designed to sit at or just below grade level, so as not to provide an impediment to traffic. Some are even topped with a small layer of paving material for uniformity. Manhole covers may have a variety of patterns and designs that can be unique to the municipality or foundry of construction. Holes may be present for ventilation (especially on stormwater systems). Special recesses to allow for pick access and removal are often present. These pick holes may or may not penetrate through the cover.

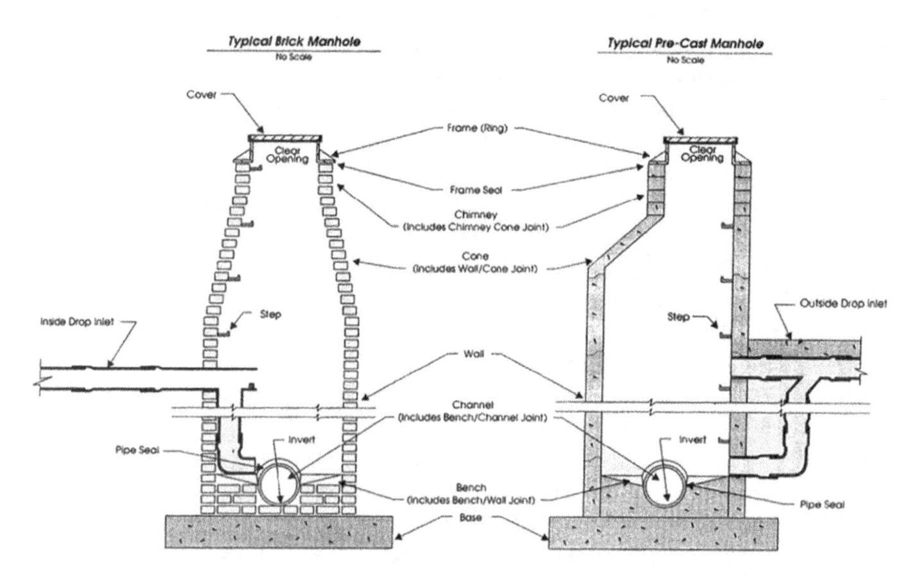

FIGURE 9.2 Diagram of major components of brick and precast manholes. Not all manholes will have all components, but most follow this general structure. Defects can occur in any section and should be noted during inspection. Image taken from Kaddoura and Zayed, *Tunneling and Underground Space Technology*. STM (Elsevier): https://www.sciencedirect.com/science/article/pii/S0886779818301974#b0070

Manholes can also indicate the type of system they are attached to on the cover itself. Some have the words "sanitary" or "gas" embedded in the cast to show what type of utility lies beneath. Others abbreviate using an "S" to indicate that a manhole is part of a sewer system. Telecommunications manholes may have the word AT&T or the Bell System logo emblazoned somewhere on the cover itself. This may seem redundant, as a specific utility may only have one manhole in a location – but this functional branding makes it easy for crews looking at three similar manholes in an intersection to decide which one should be accessed for inspection or repair work.

In the United States, nearly all manhole covers are round – primarily for safety reasons. Round manhole covers cannot be reoriented to fall into a round opening. Square and rectangular covers could fall through along the longer diagonal axis. However, while the shape of covers has become standardized, there is no single standard size for a manhole cover. For example, Neenah Foundry, a Wisconsin-based producer of manhole covers, produces lids with diameters ranging from 23 1/4″ to 26 5/8″, and thicknesses ranging from 3/4″ to 1 1/2″. While that range may seem small, it is significant, as manhole covers cannot always be easily swapped, or mixed and matched (Figure 9.3). Further, historical manholes can show even more variance – diameters as small as 20″ are possible. If an inspection contractor says that their technology requires a 22″ manhole opening for use, don't just assume that your manholes will be fine because they are "standard". Measure and confirm to avoid paying for emergency construction and crew standby rates.

Iron manhole covers are typically cast. Grey iron offers a mixture of high strength and corrosion resistance. Ductile iron can be used in applications requiring higher compressive strength. Steel is not commonly used due to its tendency to corrode, but

FIGURE 9.3 Manhole covers come in all shapes and sizes, as this collection of scrapped devices illustrates. These are from a single section of a road and its sidewalk that are being repaved. Note there is no single standard manhole opening. Image provided by W. Carter. Creative Commons: https://commons.wikimedia.org/wiki/File:Scrapped_manholes_and_covers_in_Brastad.jpg

it can be found in stormwater gratings and other less-dense structures. While these materials are impressively strong, remember that they can crack – easily, especially if dropped on edge or on a sharp material that can provide a nucleation site for a crack. Be careful when removing a manhole cover for access or inspection – cracked manholes cannot be safely reinstalled, and if a spare is not available, barriers to prevent public access may need to be quickly erected.

Regardless of the actual dimensions, manhole covers are heavy, weighing in excess of 200 or even 250 lbs. Because covers contain a significant amount of metal, theft can be a serious problem, especially in times where scrap metal prices rise. Some municipalities may bolt (using security heads) or weld manholes to the frame to deter casual theft. In other cases, bolting or welding can be done to ensure a water-tight seal, to prevent discharges and overflows, or as security provisions. Before presidential visits, the U.S. Secret Service has been known to weld manhole lids to prevent underground threats in areas where a motorcade is scheduled to pass through. In a more extreme example, manhole covers are also welded shut before some street races (such as Formula 1), as the aerodynamic characteristics of the cars could provide a sufficient force to cause a cover to lift off while a car is driving on top of the structure. While many dispute whether or not aerodynamic forces alone would be strong enough to remove a manhole cover, at least one race car has burst into flames after it was struck by a manhole cover dislodged by another vehicle – thus, welding before street races has become a common practice.

Manhole covers are often mated with a frame – a round structure with a recessed lid that the cover can lie in. Frames are intended to be attached to the manhole stack more permanently, and remain in place while the cover is removed and reattached. Frames are typically made of the same material as the cover and are often produced by the same manufacturer to ensure fit. Frames can also crack or become dislodged – presenting a serious safety concern. While a manhole cover is designed to not fall through the frame it lies in, it could easily fall through the underlying cone, injuring or killing a worker below. A loose frame can also move in a way that presents a hazard to foot or vehicular traffic.

Frames may be attached to one or more rings, intended to lift the cover to grade level. Standard rings may be constructed of concrete and allow for a longer connection to the manhole cone, forming a chimney. Rings can be installed with gaskets to prevent water intrusion at points of assembly. Adjusting rings serve a slightly different function. Over time, repaving streets may result in a situation where multiple layers of asphalt are much higher than the original roadbed, and a manhole installed in the street must be lifted for traffic flow. Rather than dig up the frame and install a traditional ring, adjusting riser rings sit in the original frame and allow the cover to be elevated by a few inches. These risers are typically constructed of cast iron and can be stacked multiple times to allow for a greater rise. Iron risers can crack or break, so inspect them carefully to ensure structural integrity.

While convenient, rings provide an opportunity for water to intrude into a manhole structure. If the seal between rings (or any other vertically stacked manhole component) is not tight, groundwater may have an opportunity to flow into the manhole. This can be especially true if surcharge conditions damage or dislodge components. This infiltration can be difficult to notice during dry weather, and it is often recommended that manholes be inspected during wet weather events to check for defects that may permit water intrusion.

Sometimes a chimney is constructed from brick, rather than rings of precast concrete. This brick structure is built up on top of the cone to the area where the frame sits. A brick chimney can be encapsulated in concrete on the exterior to form a seal. However, in many cases such structures act as a source of water infiltration, as cracks in the concrete and regions of missing mortar provide ready conduits for water to enter a manhole chamber. They are also infrequently addressed in rehabilitation programs – perhaps because they are often covered by concrete, or are relatively high components in the system. It is not uncommon to view water pouring in through these brick chimneys during a heavy rain event.

While visual inspection techniques will be discussed in more detail, it can be impossible to tell whether flaws like cracks are superficial or substantial during dry weather. Even a thorough visual inspection performed on a sunny day can fail to detect sources of infiltration. Consider temporarily installing water sensors in manholes as part of a comprehensive inspection program to detect these defects, and send crews to perform manhole inspections during the rain. Mystery sources of infiltration can often be detected when manhole chimneys receive scrutiny.

Below the chimney and rings lies the cone – the part of the manhole structure that incorporates the change in diameter from the cover at the surface to the larger structure above the sewer line. While manned entry through a 22″ portal is possible, it does not provide much room to work. The cone provides a transition from a smaller space needed to provide access to a larger space that can actually enable a worker to move and perform tasks, such as cleaning or servicing equipment. Cones may be precast or constructed from brick and provide the same opportunity for water ingress as higher structures.

Many cones do not execute this diameter change symmetrically – one side of the cone will retain a constant curvature to align with the chimney, while the other will expand accordingly. This was often done to permit the installation of steps, enabling a person to climb down into a manhole without the need for cumbersome equipment. These steps are often made of iron and are extremely dangerous as they are vulnerable to hydrogen sulfide corrosion, and can snap off when weight is applied. This can result in serious puncture wounds or even a fall into the depths of the manhole. *Never* rely on older steps or step bolts as the sole method of accessing a manhole, and *never* enter a manhole alone. Always use an appropriate team when entering a confined space. While newer steps are made of materials like steel-reinforced polymers and promise safe operation over longer lifespans, many sewer workers are injured by older devices that have failed. Use extreme caution if built-in steps are part of an access plan.

The cone sits atop one or more barrel sections (sometimes referred to as the wall of the manhole) that extends the large diameter access structure down to the bottom of the manhole. These barrels are typically 4′ in diameter and provide ample room for activity. The bottom section of the barrel can be structured to contain the pipeline itself – typically a half-pipe opened to the top for access. This region is known as the channel, and in many cases, does not take up the full space of the manhole, especially when comprised of smaller 8–12″ lines. Surrounding the channel is a sloped structure called the bench, which is intended to direct any ancillary flows into the channel – such as water entering through the manhole, or a decline in flows after a surge. Benches can be constructed from concrete or brick. Be careful – while many work on benches, sewage and debris can leave them slippery and they can present a fall hazard, especially when wet.

Many manholes are constructed over areas where the direction of flow changes (Figure 9.4). These bends can provide natural locations for problems, as debris seldom navigates contours as well as effluent. Constructing manholes in those locations provides easy access to problem locations for cleanout and repair. Similarly, manholes are often located at the intersection of multiple lines – where more than one inlet will enter such that flow converges into a single outlet. Some of these inlets will even come in at different elevations, leading to drop connections within the manhole itself. While collocating manholes with each of these features can be logical for facilitating maintenance, remember that any disruption to laminar flow provides an opportunity for the release of hydrogen sulfide. Use extra caution in manholes with these features, and consider installing hydrogen sulfide monitors below covers to develop a longer-term model of gas buildup.

As noted, manholes come in different shapes, sizes, and types. This description of components applies to a typical manhole installed in a wastewater system – manholes for other purposes can diverge in a variety of ways. For example, a manhole constructed to access steam tunnels may be bolted, hinged, fitted with a retractable ladder, and open directly into the tunnels themselves. Other manholes may be constructed

FIGURE 9.4 Close-up of the bench and invert of a manhole. For MACP purposes, the outgoing pipe connection will always be indexed to 6 o'clock. There may be multiple incoming connections, so they are not desirable frames of reference. Image provided by Pam Broviak. Creative Commons: https://ccsearch.creativecommons.org/photos/63aa2626-7e3b-442c-bb47-873887b8a67b

atop access shafts used to enter large diameter interceptors. These structures may consist of long sections of barrel that extend down 80–100′, sometimes bored through solid rock, and are free of ladders or other structures to facilitate access.

To establish some kind of order to this variety, building codes and other specifications will sometimes group manholes based on depth, rather than function. Shallow manholes or access chambers are typically less than 3′ deep. These manholes may be placed at the start of a sewer line, or directly atop a communications vault or steam tunnel system. "Normal" manholes are approximately 5′ deep, and often consist of the structure and components outlined above. They may or may not be equipped with stairs or a ladder. Deep manholes are anything beyond this, and are often constructed to be wider in diameter (in the barrel) than other, shallower manholes.

CONDITION ASSESSMENT APPROACHES

Many buried infrastructure systems are constructed in phases – sometimes over many decades, by a variety of contractors. Private systems may have undergone multiple changes in ownership – telecommunications manholes may have started as AT&T assets, shifted to the Bell regional operating companies, before coming to reside with Verizon or other modern successors. As-built records may be in various stages of completeness, and the surface topography may have changed significantly. A structure that was constructed in the shoulder of a road may now be in the center of a highway. Further, much of this construction occurred in an era before computers, and when establishing a precise location required an expensive survey.

It should come as no surprise to discover that many system operators do not have comprehensive knowledge of exactly where all of their manholes are, and many condition assessment programs begin with a thorough locating project – often using Global Positioning System (GPS) receivers. In these programs, crews are provided a list of manholes, best available maps, and are tasked with finding and recording an accurate location of assets. Some programs also couple this with a project to excavate and raise manholes that may have been covered or buried over time (Figure 9.5).

GPS receivers are key to this task. GPS is a system of more than 30 satellites currently in orbit around the earth. Each satellite continuously broadcasts information about its location in orbit and a precise measurement of synchronized time (using an atomic clock). Devices on Earth detect these signals and can measure the difference in time between the signal's transmission and receipt in order to calculate how far away the device is from the satellite. Performing these calculations on broadcasts from multiple satellites can enable a device to establish a precise measurement of its location on the surface of the Earth through triangulation. Positional accuracy increases with the number of satellites used to determine the measurement.

Historically, the United States government would intentionally degrade signals broadcast by GPS satellites – in order to prevent foreign adversaries from using the system to gain an advantage in times of war. Although the government no longer does this, consumer-grade GPS devices are still unable to generate precise location information without employing some form of (expensive) data correction. As

FIGURE 9.5 Failure to raise manholes to street grade can cause a number of problems, including, but not limited to, infiltration. Many manholes are simply paved-over and can be difficult to locate in subsequent years. Image provided by dwfree1967. Creative Commons: https://ccsearch.creativecommons.org/photos/ecbf6058-8da6-4eaa-a2a1-b5d0ff3ce7f5

a result, the typical GPS receiver in a consumer device may have location errors of more than 30′, due to distortions introduced by the ionosphere or nearby structures. Anyone who has ever seen their location on a smartphone GPS system jump from an interstate to a side street and back while driving can confirm this lack of precision. If these consumer devices were used to locate manholes, the results could be of limited utility to crews – a 30′ error could put a manhole in the woods, on a different street, or in a very different location than would otherwise be expected.

Survey-grade GPS systems employ different types of corrective algorithms to provide more precise location information. Some units have onboard clocks that remain synchronized with the satellites, while others employ algorithms like real-time kinematics (RTK), which examine the phase of received signals in order to determine propagation delays and other sources of errors. Some receivers intended for international use incorporate data from other satellite clusters, such as the Russian GLONASS system, in order to accurately calculate position information. Regardless of the approach used to correct the satellite signal, so-called "survey-grade" GPS devices are widely available (Figure 9.6).

Many of these devices utilize external antennas in order to filter satellite data and generate a precise location estimate. Units manufactured by Trimble have become synonymous with corrected GPS, though devices from Leica and Topcon can provide comparable performance. Professional units from Garmin and Navtech continue to improve, and system integrators in China are beginning to incorporate RTK corrections into no-name GPS units. This could dramatically drive down equipment cost in the near future. While these units may not be suitable for true survey work, they may provide accuracy that is good enough for field manhole location.

While these GPS receivers may be colloquially called "survey-grade", they are actually rated in terms of their positional accuracy. Submeter GPS systems are able to locate a position on the earth's surface with an error of less than a meter. Submillimeter systems cost an order of magnitude more, but are able to provide positional information that is much more accurate. For manhole location surveys, submeter accuracy should be the requirement specified for inspection crews. If equipment fails, do not allow the resourceful to complete inspections using an iPhone or Android device. Stop inspecting and replace a failed submeter GPS receiver with one of equivalent functionality. Erroneous or inaccurate consumer GPS readings can introduce issues in location surveys that can plague municipalities for years to come.

Even knowing this, it can still be tempting for some to use a consumer device when equipment is not working properly. When specifying standards for GPS inspection, in addition to specifying the placement of the receiver (center of the manhole cover is common), also consider specifying both a receiver data file and a screenshot of the display. It is also good practice to require that crews take photos of the manholes in

FIGURE 9.6 Survey-grade GPS units are often classified by their resolution – submeter and subcentimeter units are popular for manhole locating work. Notice that they look very different from traditional, consumer-grade units, and often attach to external antennas or pointers for added accuracy. Photo provided by Harri Blomberg. Creative Commons: https://commons.wikimedia.org/wiki/File:GPS-m%C3%A4tning_p%C3%A5_j%C3%A4rn%C3%A5ldersboplats_i_Ytterby_sn,_Bohusl%C3%A4n,_den_28_april_2006._Bild_2.JPG

context – if there is any issue with the GPS data in the future, these photos can help locate the manhole by showing it in its physical surroundings.

Further, photos can reveal issues with the manhole that may not be detected by GPS or a subsequent condition assessment. Many manholes are not regularly raised every time road construction or resurfacing work occurs. This can lead to sunken structures that are slightly below grade. While this can be an inconvenience to passing traffic, it can also transform an imperfectly sealing manhole cover (or one with ventilation holes) into a major source of infiltration as it becomes the low point in the road. Photos can clearly show this situation in context. This can be critical, as these sources of infiltration can be quickly and inexpensively addressed through the installation of devices like dishes that sit just below the manhole cover.

It can be tempting to pair GPS data collection with manhole inspections, but this can cause efficiency to suffer. It is reasonable to task the locating crew with some condition assessment tasks – maybe describing the condition of the cover and determining if it is bolted or welded – in addition to GPS location and contextual photography. These tasks can often be completed by a crew that can move quickly from manhole to manhole, with minimal traffic control. They can also note manholes that will require work prior to an internal inspection – such as those that will need to be uncovered or have welds ground open. The follow-on inspection crew will have a much easier (and more productive) time if these tasks are taken care of prior to their arrival on site.

After manholes are located, the next step is often condition assessment. Fortunately, NASSCO's Manhole Assessment and Certification Program (MACP) standard has created a framework that describes the types of information to collect, as well as standardized ways of describing defects. This framework is intended to be followed by crews making physical measurements and is divided into two levels. Level 1 inspections are more basic, cursory inspections that provide an assessment of the general, overall condition of a manhole – all of the information required for a Level 1 inspection can be completed without a manned entry. Level 2 inspections are much more detailed and obtain enough information to identify the general condition of a manhole, measure its dimensions catalog any defects, and provide enough information to repair or rehabilitate the structure. This comprehensive picture of a manhole requires entry in some form – either with measurement equipment, a camera device, or human inspector.

More specifically, a Level 1 inspection requires completing various fields of the "header" information on an MACP inspection form. Just as was the case for PACP, MACP inspections are structured in a way that allows them to be completed via a paper form, even if most modern inspectors use some kind of software to assist. Level 1 fields include the date of inspection, the manhole number and location, and general notes about the condition of the cover, rings, frame, chimney, cone, barrel (wall), bench, and channel. Specified terms are to be used in these condition descriptions in order to ensure standardized results. For example, covers may be listed as missing, cracked, corroded, broken, or sound, while chimneys and barrels may be noted as defective (D), sound (S), or unknown (X), with details noted in a separate field.

A Level 2 inspection requires collection of all of the Level 1 information, with additional detail required. Notably, this includes several measurements: rim to invert, rim to grade, and grade to invert. It also requires much more detailed information about all manhole components, including material information, component depth measurements, and a list of all pipe connections with clock positions. Level 2 inspections also require interior steps to be counted and evidence of surcharge to be noted on the inspection form.

Level 2 inspections also require defects to be observed and coded using codes that are derived from and are similar to PACP codes. These codes typically have two parts – the first code represents the manhole component, and the second part represents the defect. For example, a section of the manhole chimney that is missing a brick would be coded as CMI for chimney interior followed by MB for missing brick. The missing brick code is the same in PACP and MACP inspections, minimizing the likelihood of errors. Defects must be described in their clock position within the manhole, as well as the depth from the cover. Clock positions in a manhole are derived from the outgoing pipe, which is defined as the 6 o'clock position.

Continuous defects are treated differently in MACP. Defects in manholes must be truly continuous to be noted. Further, continuous defects may not extend through multiple types of manhole components. For example, if roots enter at the chimney and continue into the cone, the defect must be closed at the end of the chimney and reopened as a new defect at the cone. In order to be continuous, defects need only extend for one vertical foot in a manhole inspection.

Remember that while MACP requires that defects be identified and coded, it can be extremely difficult to determine whether or not defects are sources of infiltration in a dry weather event. A good practice is to flag manholes with certain types of defects – missing bricks, large cracks, internal offsets – and schedule follow-up inspections of these manholes during rain events. This can give a contractor something useful to do when wet weather makes pipe inspections difficult, or can be a good task to assign to internal crews.

The information listed in this section in the description of Level 1 and Level 2 inspections is representative only. It is not intended to be an exhaustive list of the detail that must be recorded in an MACP inspection. For a full list, refer to NASSCO's PACP/MACP certification materials. However, some key items that are omitted by the standard should be noted. MACP recommends that inspectors take contextual photographs and provides locations to enter GPS coordinates. It does not actually require that this information be obtained in an inspection. If an asset owner wants to ensure that GPS coordinates are obtained and that photographs are taken for each inspection, these items must be specified separately in a contract specification. It is not enough to say MACP Level 2 and assume that they will be included.

While the levels provide a convenient way to specify the level of detail required in a manhole inspection, they are not suitable for all municipalities. A Level 2 inspection can cost more than twice as much as a Level 1 inspection, and may provide more information than a municipality is looking for. Thus, "Level 1.5" inspections have emerged. These inspections are not a part of an official NASSCO standard, and include all of the features from a Level 1 inspection along with a subset of features from a Level 2 inspection. These inspections typically cost less than full Level 2 inspections and can be completed much more quickly. For example, it is common for municipalities to specify Level 1 inspections with rim-to-invert measurements for each structure that could be accessed.

Interestingly, some information on the MACP inspection form does not need to be filled in for a Level 1 inspection, but can be readily noted. For example, evidence of surcharge is a field that is only required of a Level 2 inspection. However, if a manhole lid is removed (as it should be to note chimney, barrel, and bench condition), and ragging is visible on the steps, best practice would be to note this on the form, even if it is not required. While the flexibility to note additional information and perform Level 1.5 inspections is easily accomplished on paper, many software packages have yet to enable this functionality. In order to ensure that operators fill out required fields and minimize errors, many programs are preset for MACP Level 1 functionality and MACP Level 2 functionality – it is difficult to manually enable additional fields to create a custom inspection. Hopefully, this will improve with time, as municipalities remain slow to embrace full Level 2 inspections.

Robots and other tools can be used to capture video that can supplement or assist in the defect coding of an MACP inspection. For example, lowering a robot equipped with a fisheye lens into a manhole can capture its entire interior so that defects can be coded by dedicated staff after the deployment. These robots are not required as part of an MACP inspection, but can provide a value-add in the form of inspection videos and the ability to look at a manhole in more detail at a later date. Some vendors use the same devices employed in pipe inspections. Both the Ibak Panoramo and RedZone Solo can be used to perform manhole inspections and pipe inspections, using their fisheye lenses. Lighting can be a challenge, especially for cameras that automatically adjust brightness and contrast. Tenting may be required to produce videos that are lit appropriately. Remember, many of these robots were designed to inspect 8″ pipelines, so lighting up a 4′ barrel may require some adjustment to parameters. Always ensure that manhole inspection videos have proper lighting prior to acceptance.

Other devices were created for the sole purpose of inspecting manholes, including the Cues Spider and Envirosight CleverScan. These devices typically combine a tripod and inspection unit into a device that can be lowered into manholes with a simple button press. While some units may rely on a single fisheye lens in order to capture data, other units employ multiple high-resolution cameras and stitch the individual images together in software to create a continuous panoramic image. This approach tends to produce a more detailed image than fisheye cameras (regardless of resolution), as data is more evenly distributed by the lenses onto multiple CCDs, and one imager is not attempting to capture data for the entire interior structure.

These devices can also construct a 3D point cloud of the interior of a manhole using a variety of technologies. Some units use lasers or lidar units which function similarly to those used in pipeline inspections. Others use stereocameras to generate the same type of data via a different mechanism. Stereocameras are devices that use more than one camera mounted at a fixed offset in order to make distance measurements. Human eyes operate according to a similar principle. Objects located far away from us tend to be centrally located in each eyeball. However, as objects move closer their relative position within the eyes changes, gradually moving to the interior of each eyeball (closer to the nose). Our brain processes this change in order to provide an accurate depth measurement – for a sense of just how accurate, close one eye and try to grab a pencil held in one hand with the other.

The same is true for stereocameras, except in this case, an onboard computer translates the overlap in individual images into a distance measurement. When the same feature is detected by more than one camera, the position of the feature in each camera can be used to make a distance measurement. As multiple images are captured, this same principle can be used to measure the distance traveled – a concept known as visual odometry. Stereocameras operate most effectively when they are placed in feature-rich environments – where discrete items can be clearly noted in each camera, and the item is distinct enough from its surroundings in order to be clearly detected. Using a stereo camera in a featureless environment, such as smooth pipe, can lead to a concept known as slippage, where features tend to stretch out in size as the stereo camera struggles to identify the appropriate offset.

Regardless of approach, point clouds generated in manholes can contain sources of error. While vendor animations show smooth deployment, a degree of twisting and turning on the cable tends to occur frequently during these inspections, making it difficult to reliably say (in the post-processing stage) whether a crack is truly at 12 o'clock. Water droplets appearing on the lenses or lasers can also create significant voids in point clouds. However, these errors tend to be much less significant, as point clouds are not used to establish the structural integrity of manholes. Instead, they tend to be used to measure the height of drop connections and estimate the interior volume for lining applications. Errors in measurement of an inch tend to have a much less significant impact on rehabilitation work.

One major drawback to these automated devices is their ability to collect rim-to-invert measurements. While a human at the surface can easily obtain this information with a measuring stick, electronic devices are forced to approximate this measurement. Some devices use laser rangefinders to measure the distance to the bottom of a manhole. These devices cannot penetrate the flow and will produce a measurement that is a bit less than the true dimension. Some devices use an extension rod at the end to measure this distance. By lowering the device to the bottom of the manhole and adding the length of the device, extension rod, and amount of tether deployed, the rim-to-invert distance can be calculated. With slow flows, this approach can be effective, but rapid flows tend to pull the extension rod downstream when contact is made, adding a few inches to the depth measurement. While these errors can seem insignificant, when input into a hydraulic model they can be pronounced. Fortunately, they can be easy to spot, as gravity sewers seldom flow upstream from manhole to manhole.

One issue that often arises in the course of manhole inspections is data management. When performing a pipe inspection, contractors need only keep track of the CCTV file that the robot obtains during traversal. Manhole inspections often require the collection of video data, photos, notes, and GPS coordinates – each of which needs to be tied to a specific manhole. Errors in classifying and assigning files can easily occur – for example, if crews take contextual photos with a digital camera and accidentally copy them into the wrong folder. These issues are inevitable, but operational controls can help prevent or minimize their occurrence. For example, some municipalities require crews to spray paint the manhole number on the bottom of a cover, and include that in one of the photos. This way, if a photo of a different manhole appears in an inspection, it will be obvious that an error occurred. Use special care when relying on standalone digital cameras. When the battery in many of these devices loses its charge or is removed, the internal clock can be reset – leading to a slew of manhole photos obtained on 1/1/1999 (or whenever the reset date is for that manufacturer). When data consists only of similar-looking manhole photos (with no usable timestamp) and header forms, it can difficult to detect mistakes until a subsequent inspection.

While much effort focuses on inspecting and assessing the condition of manholes as standalone units, manholes can also be useful indicators of problems elsewhere in the collection system. Surcharged pipe conditions can manifest themselves dramatically in manholes. Oftentimes, when a manhole cover cannot be immediately located, a quick scan of the surrounding area will reveal it to be lying several feet away – as it may have been blown off during a surcharge. While this can represent a dangerous situation, and many municipalities weld or bold manhole covers in place for this reason, tracking the number of times wastewater reaches a manhole cover can provide useful information about the flow within the system. As a result, a variety of manhole-mounted sensors have become popular.

Using manholes as sensor locations is a relatively novel idea, tied closely to the availability of low-cost, digital cellular and satellite communication. Whereas previously, sensors would have needed to be wired into an expensive grid of power and data connections, many sensors now come with embedded batteries and wireless modems that allow them to be installed in any location making it easy to create a wireless network of sensors. Some of these sensors are designed to detect tampering with the manholes themselves, while others are used to obtain information about the flows beneath, including levels and quantity, as well as information about the composition of the fluid being transported.

Tamper-detection sensors initially saw widespread use in locations where high scrap metal prices made manhole covers attractive targets for thieves – many cities in China drove the development of these devices. However, their usage has continued to spread, due in part to the fact that sewer overflows and other phenomena can displace manholes without any intervention from people. If left uncorrected, these events can create serious threats to human safety – a misstep into an open manhole can cause injury, or even death.

Many of these devices use simple contact or tilt sensors. If a manhole is opened, the sensor straightens out or loses contact and the device automatically signals that a tamper event has occurred. This event could be an overflow, theft, or other activity. Many sensors are mounted just below the frame and contain internal antennas for connection to a cellular network. Others have built-in low-powered radios, and interface with a wireless gateway or external system for long-distance signal transmission. Standalone tamper sensors are available from companies like Gemtek and

WiiHey. Other more complex sensing devices include tamper detection as part of their standard functionality.

Other sensors are designed to detect characteristics of the flow below manholes. SmartCover provides a series of ultrasonic sensors that attach to manhole covers and can monitor properties of the flows passing beneath. The SmartLevel measures the height of water below a manhole cover, records this data onto internal storage, and sends this data to a cloud infrastructure platform through a satellite communications system. Other devices like the SmartFLOE can actually estimate flow rates and transmit this information to the same cloud system. While beyond the scope of this text, it is worth noting that the SmartFLOE relies on Manning's Equation to estimate flow and requires accurate roughness coefficient data in order to produce valid estimates. While some roughness coefficients can be looked up in reference tables, more optimal values can be obtained from field measurements. Level sensors from SmartCover also include the UnderCover theft detection sensor, which uses an accelerometer to detect whether or not a cover has been disturbed.

The key benefit to level monitoring devices like the SmartCover is the ability to detect conditions that could lead to a CSO or SSO before they actually occur. By setting a threshold appropriately, users of these level sensors could configure the system so that crews are paged or alerted with enough time to respond and prevent the overflow. In one published case study, a crew was able to respond to a water level alarm, detect a surcharge as it was occurring, and employ a vacuum truck to remove a blockage before the system ever overflowed.

Further, these devices can be installed on an as-needed basis, and do not require the installation of a broader platform in order to operate successfully. Individual SmartCovers can be ordered and deployed to manholes that are known to be at risk of overflows or intrusion. Additional units can be added as time and funds allow. This makes it possible to gradually build up a network of flow monitors over time, and establish a hydraulic snapshot of an entire collection system, at a cost that is well below that of a traditional wired setup.

The downside to systems like the SmartCover is that the sensors and analytic suite are proprietary. Once a municipality starts using SmartCovers, it can be difficult to switch to or add on sensors from a different vendor in the future. This also means that bids for new equipment tend to come from the same regional distributors, meaning competition does not effectively drive down the prices of these units.

Other sensors can be suspended below a manhole cover in order to obtain information about a wastewater collection system. The same hydrogen sulfide meters that have become a staple of confined space monitoring and were discussed in a previous chapter can be hung inside a manhole to log data on gas levels over time.

Many public health or epidemiological studies have noted that valuable information could be obtained from studying the waste products of a population. Data about disease, recovery, drug use, and eating habits can all be decoded from excrement. Biobot Analytics has developed a series of sensors that are submerged in a stream of sewage in order to draw conclusions about the population producing said sewage. Their sensors can be mounted on the underside of manhole covers – a critical factor as the components they are detecting break down by the time they reach the wastewater treatment plant.

While these sensors are of limited utility to wastewater treatment plant operators, they can draw valuable conclusions about the health and wellness of the surrounding population. For example, in 2020, Biobot used its sensors to analyze residential sewage for COVID-19 titers (biomarkers produced in response to an infection) and concluded that cases of coronavirus were much more prevalent than official statistics suggested. This enabled public health authorities to shape testing and treatment programs in parts of Massachusetts where the samples were collected. There is some risk to deploying dangling sensors like the Biobot – if the device got loose, it could create a blockage elsewhere in the system, or potentially damage treatment and filtration equipment. However, these risks further highlight the potential for collaboration and partnerships between collection system operators, community organizations, and startup companies. By working with wastewater professionals, the Biobot team could design their device to withstand typical sewage conditions. By working with the Biobot team, the collection system staff can generate positive press and beneficial results for the communities they serve.

An additional problem that can impact a variety of manholes is especially prevalent in areas that have buried power infrastructure. Underground power lines tend to be much more problem-free than those mounted on utility poles. While the latter can be impacted by wind, trees, and even people, the former tends to be buried securely in underground vaults. Unfortunately, age, vibration, and water intrusion can cause the insulation shielding buried power lines to crack and decompose, and wires coming in contact with adjacent surfaces can cause undesired electrification. When these electrified materials extend to the surface, the result is termed a stray voltage – contact with which can seriously injure or kill passersby.

While the statistical chances of this happening at any one manhole may be slim, power companies with burned infrastructure have hundreds of thousands or even millions of pieces of equipment that could fail within a single city. In New York City alone, more than a thousand sources of stray voltage hospitalized two pedestrians and killed multiple dogs in a single year. This resulted in the death of an individual who stepped and fell on an electrified manhole cover on the sidewalk.

While these stray voltage events are likely foreign to the typical wastewater system operator (most wiring in sewer systems is low voltage), they represent one of the most significant hazards that power distribution companies must deal with. Addressing stray voltage requires both detection and remediation.

Detection of stray voltage can be extremely challenging, as it is not a continuous condition. Just as some sources of infiltration only manifest themselves during rain, some sources of stray voltage only come in contact with the surface during times of heavy load, high temperature, or precipitation. Detection of stray voltage can occur through testing that occurs in close proximity or at a distance through the use of various sensors. Close proximity testing tends to be the simplest. In this case, an individual brings a tester "pen" next to an object that is suspected to carry an alternating current. If such a current is present, it will induce a capacitance in the device and cause a light or buzzer to sound. Other contact-devices, such as voltmeters equipped with shunt resistors, can also be used to confirm the measurement.

However, this process requires some informed selectivity on the part of the tester: an inspection crew must knowingly identify and test an object that is suspected to be the source of stray voltage. Innocuous objects that are not intended to carry dangerous electrical currents will often not be tested – for example, improper demolition of buildings can cause piping or other metallic materials that typically would not be examined to become energized. In order to detect these phenomena, electric field detectors are often employed. These detectors can be mounted on a car, truck, or other mobile platform and move around an area, scanning for electric fields. Such instruments respond to any electric field that is present in an area, making them prone to false positives that can require the use of many close-proximity measurements to rule out various pieces of equipment.

Both of these inspection approaches are limited in terms of time – they only scan manholes in a system for a brief moment, meaning an erratic stray voltage may be difficult to detect. Some sensors like the Sentir, by CNIguard, can be mounted under a manhole and attached with a length of wire. If the manhole becomes electrified, the Sentir will signal this result through a cellular or radio signal, and crews can be dispatched to remediate. Since the Sentir is permanently installed, it can monitor structures for a long period of time – meaning that if a stray voltage only occurs during rain or periods of high load, the sensor will remain in place long enough to detect the issue.

Fixing stray voltages can require an intensive effort to overhaul manhole infrastructure. Some of the simplest solutions involve installing an insulator between the manhole and the rest of the buried chamber. While easy to retrofit, these insulators can fail or be compromised when excessive rainfall occurs – times in which individuals are most likely to slip and fall and come in contact with the surface. This isn't to say that insulators aren't a good idea – they are a great first step in resolving stray voltages, but a true solution to the problem usually involves finding the underlying fault in the electrical infrastructure.

Finally, a note is in order for those who are skeptical of dedicated manhole inspection programs. Many people who administer buried utilities find it difficult to believe that manholes could be the source of significant problems in a system – after all, these structures are accessed every day, and crews interact closely with manholes when they are upgrading, repairing, and working with a system. The conventional wisdom supposes that serious defects would be noticed and reported in the course of normal operations.

While there is a degree of logic to this thinking, it is not borne out by reality. It is true that crews interact with manholes on a daily basis, but mostly in the course of accomplishing a larger task – removing a blockage or deploying an inspection camera. It follows that personnel are focused on the task at hand, such as ensuring that the camera does not get damaged during insertion into the system. Small defects such as missing bricks or cracks in the cone may not be noticed peripherally. Even if they are noticed, without a system to immediately record and log them, they are quickly forgotten.

Even though we all drive vehicles every day, it should come as no surprise that a regular inspection turns up problems that we didn't notice during the course of commuting to work or hauling equipment. Even if the car idles strangely when cold, by the time we arrive at our destination, we've forgotten that the problem even existed in the first place. Crews use manholes just as the rest of us use vehicles – as tools to accomplish a task. Specialized inspections are needed to identify problems in these tools before they degrade to the point of serious failure.

LATERALS

While manholes are a significant category of often-overlooked structures that can contribute to problems in a buried utility, they are by no means the only structure that fits this description. Private sewer laterals (PSLs) represent another commonly overlooked structure that can contribute significantly to problems in wastewater collection systems. Unlike manholes, which can be found in a variety of utility networks, lateral lines are restricted to sewer systems. However, they have only recently become part of comprehensive condition assessment programs.

When municipalities and sewer authorities describe a collection system, most are referring to the network of publicly owned pipelines – ranging from smaller 6–8″ mains, to much larger interceptors or trunk lines. These assets are clearly public responsibilities – they were constructed by government programs, they often lie underneath public streets, and they aggregate and transport wastewater from multiple property owners to a wastewater treatment plant for processing and subsequent discharge. Yet, many municipalities struggling with I/I problems may find that even when this portion of the collection system has been exhaustively inspected and overhauled, issues still exist.

In many cases, the reason can be traced back to PSLs, often simply referred to as laterals. Lateral lines are the structures that connect residences and businesses to the sewer system. Lines connecting residences tend to be 4–6″ in diameter, while those connecting bigger commercial buildings will be much larger. Most modern lateral lines are constructed from PVC, cast iron, or HDPE, but vitrified clay can still be found in many older installations.

Lateral lines can be broken into two components: the upper portion of the lateral, which runs completely underneath private property, and the lower portion that may pass under a sidewalk or other public property in order to connect into the public sewer. The demarcation between these points may be established by the property line or the presence of a cleanout. However, in many municipalities, the entire lateral is the responsibility of the property owner – even the portion that passes under a sidewalk and connects to the sewer. Some sewer authorities maintain responsibility for the lower portion and leave the upper portion to the property owner, but these jurisdictions are in the minority.

The goals of property owners and municipalities with regard to lateral lines tend to be very different. Property owners maintain their lateral lines to the point where sewage is conveyed off of the property. This means that they place a great emphasis on removing obstructions like roots and grease that would cause the line to back up into their residence or business. Sources of infiltration have no impact on their usage of the line, and property owners have no incentive to proactively address them. Thus, if roots enter through a joint in the lateral and begin blocking the flow, the property owner may cut down an offending tree and hire a plumber to cut out the roots. However, they are unlikely to engage in further rehabilitation – owners are unlikely to repair any offset joint caused by the roots that may allow infiltration into the lateral.

While this may not seem significant, on a national scale, the American Society of Civil Engineers estimates that 500,000 miles of PSLs exist to feed into the 800,000 miles of public sewer lines in the United States. In an individual system, the EPA estimates that private laterals account for almost 50% of the length of a collection system, and contribute more than 40% of the infiltration and inflow into that system. In some cases, this I/I comes from illegal connections like roof drains and downspouts. However, other cases stem directly from degradation and defects in the laterals themselves. It follows that addressing infiltration problems or coming up with a comprehensive I/I strategy must incorporate lateral condition assessment in order to be successful.

While the public/private boundary may have been a useful method to segment responsibility in the past, repairing private lateral lines can have measurable impact on the public good, especially if quantitatively measured against the cost of treating 1,000 gallons of flow or in fines, penalties, and cleanup costs for each SSO event that is eliminated. Even when the financial incentive is clear, legal and cost concerns can still present challenges to implementing condition assessment and repair programs for laterals.

Since laterals are simply smaller versions of sewer lines, many of the same condition assessment technologies can apply. Smoke testing is commonly used to evaluate the main line and associated laterals. In smoke testing, blowers are used to force smoke into an isolated section of the sewer line. This smoke will spread out through the system and escape through any cracks or defects and make its way to the surface. Smoke coming out of manholes and roof vents is a good sign – smoke emerging from sidewalks, front yards, and pavements indicates the probable location of a defect allowing smoke to escape and infiltration to enter during wet weather.

Smoke tests are easy and inexpensive to perform. All that is required is smoke to inject into a system, and a crew to scan the area to look for plumes exiting the system. Many departments use the same "smoke bombs" used in pyrotechnic equipment. These devices are derived from smoke screens used by the military and generate large quantities of smoke in a short time. Unfortunately, the smoke generated by these devices contains fine particles of zinc chloride (as well as smaller amounts of chlorinate compounds). Sustained exposure can cause burning of airway passages and can lead to acute respiratory distress. While most municipalities have moved on to safer materials, the use of zinc chloride still persists, perhaps due to existing stockpiles. Anyone coming into contact with smoke from these devices should wear full respirators for protection.

Continuing to use zinc chloride is not recommended. Safer alternatives such as Hurco Technologies's LiquiSmoke are widely available, and can even be cheaper than their more toxic counterparts. LiquiSmoke is a proprietary substance that is classified as a hydrotreated middle distillate – a compound derived from petroleum. LiquiSmoke consists of a mixture of hydrocarbons and exists in liquid form at room temperature. If the liquid contacts a hot surface, it is transformed into a gaseous state characterized by plumes of white smoke. In usage, LiquidSmoke is often combined with a small gasoline motor connected to a blower system. When the motor reaches its operating temperature, pressurized LiquiSmoke is forced into the hot exhaust where it vaporizes and the resultant smoke is forced into the sewer system through the attached blower. The result is a large quantity of nontoxic smoke, though prolonged direct exposure to the vaporized substance is not recommended.

Some care must be taken when administering a smoke test – the fact that the test may result in plumes of smoke appearing in yards, basements, and other locations can easily generate public alarm. Always coordinate smoke testing with the local police and fire departments, and hang door tags on homes in the testing area 24 hours in advance. It can also be useful to include a list of frequently asked questions and a link to the material safety data sheet (MSDS) for the smoke material being used. Common questions from residents may include: will this smoke stain the interior of my home, what if I have a respiratory condition, and will pets be safe if left home during this testing. Include a number to call and a link to a website with more information, if possible.

Remember, smoke testing only indicates the presence of a defect – the actual location may be slightly off if the smoke wanders through the ground as it makes its way to the surface. It does not provide any indication of the type of defect or the circumferential location of the defect within the pipe. For example, smoke testing may reveal an offset joint in the invert of a lateral that may not contribute significantly to I/I. That said, the low cost of smoke testing, coupled with the ease of digging and repairing a section of PSL (compared to a main line or trunk sewer), makes it a valuable tool for rapidly screening large portions of the system. Crews can typically assess about 800′ of line and connected laterals every 20 minutes.

Dye tests can also be used to look for sources of inflow rather than infiltration. In dye tests, a CCTV camera is inserted into the trunk line of a sewer and advanced from tap to tap. At each tap, a fluorescent dye is poured into the downspouts of the corresponding or associated residence. If the dye emerges from the tap then it is clear that the roof drain is connected to the sanitary sewer. While modern building codes prohibit such connections, municipal regulations remain a patchwork – some locations have required homeowners to disconnect downspouts and basement sump pumps from the sewer system. Others may have grandfathered in existing connections or shifted the burden to the new owner of a property at the time of sale or transfer.

Because of this, best practice is to record all dye test results and allow office staff to address notifying property owners. There is no good reason for a field crew to get into a confrontation with a homeowner over an illegal connection. Notify them in an appropriate way, give them a chance to ask questions, and make sure all documentation is in order before requiring action. Many property owners will be unaware of this connection and will graciously fix it when notified. Others may not respond so positively. Don't let this range of human responses slow down a dye test program or create a PR nightmare. Let crews stick to testing.

However, since crews will be required to access private property to perform dye tests, many municipalities have adopted different approaches for testing. Some have crews test properties for compliance – providing notice before the testing, and pouring dye into all yard drains and downspouts. Typically, this comprehensive review will be performed once as part of an overall baseline assessment. Other municipalities make dye testing a requirement for property sale or transfer. These point-of-sale dye-testing regulations require that property owners verify the absence of illegal connections prior to closing. In many cases, the onus is placed on the current owner to hire a plumber and perform the test independent of city crews and entirely at private expense. While effective, this point-of-sale dye testing can slow the process of reducing inflow, as properties may go decades before being tested.

A variety of dyes may be used consisting of fluorescein, rhodamine, or other compounds that are widely considered to be nontoxic and biodegradable. Dyes can be obtained in a variety of colors, and some may be selected due to their ability to fluoresce when a UV light is shined upon them. This can make detection in areas of insufficient lighting much easier. It may be necessary to choose more than one color, depending on the types of properties along a sewer line. For example, inexpensive red dye may work well to test residential connections, but may not provide the needed visibility to test the drains at a meat processing facility.

Laterals can also be inspected through standard televising equipment – any homeowner who has had their sewer drains inspected can attest to this. A plumber may insert a push camera into a cleanout and look for blockages, roots, or other problems. While this approach is effective for homeowners seeking to fix problematic connections on a one-off basis, it would not be well suited to a systematic assessment of all laterals – crews would need to bring the push camera to each private property, locate the cleanout, perform the inspection, and manually move on to the next location, crossing streets and driveways as needed.

A much more efficient route to inspect laterals is from the main sewer line, and manufacturers of CCTV crawlers have developed lateral launching systems to do just this. In a lateral launching system, a CCTV crawler drives up the mainline, stopping at each tap. It centers an attachment onto the tap, orients itself properly, and discharges a camera head that functions akin to a push camera to travel up the lateral and look for defects. Launch cameras may be self-leveling or equipped with pan–tilt–zoom capabilities. Many offer built-in sonde capabilities for surface locations so that problems can be easily flagged to be repaired at a later date.

These lateral launches are often propelled by a fiberglass push rod – just like a push camera – and can travel up to 150′ into a lateral. Some camera heads can even rotate and are equipped with guide wands to enable the operator to navigate into both branches of a wye connection. Advanced units include a range of user-friendly features including lens wiping, adjustable lighting, and rearview cameras to assist with tether management during retrieval. The same considerations discussed for CCTV cameras can be used to evaluate lateral launch cameras – however, as these cameras are inspecting much smaller lines, overall resolution and image quality tends to be much less of a differentiator. Even antiquated NTSC cameras can still produce usable image quality in a 4″ line, as features are close enough to be clearly resolved.

Crawlers with built-in lateral launchers tend to be compatible with equipment from the parent manufacturer. Thus, it is easy to retrofit a Cues CCTV truck with a Cues lateral launching system, and use the existing winch, control computer, and screenwriter. Commercially available systems include the Cues LAMP II, the Ares LETS 6.0, and Ibak's LISY system. From the municipal perspective, any lateral launching system will likely be acceptable for inspection purposes – the choice of an individual system will depend much more on the needs, budget, and preferences of the contractor or crew lead.

While an I/I investigation is a significant reason for lateral inspections, it is not the only situation in which televising a line may be desired by a municipality. Utility work in the area may necessitate crossbore inspections. While directional drilling makes it easy to install gas lines or fiber-optic conduit, it also makes it easy to run new pipelines through existing PSLs. Crossbore inspections involve quickly sending a camera through a lateral to ensure that it remains intact – condition assessment details are not necessarily notated during this type of inspection. Similarly, construction work may necessitate a no-damage inspection – such as when new sidewalks are installed above a lateral. Again, a quick snapshot of a line with a CCTV camera can validate that construction did not cause any problems. To avoid argument, make sure to perform a pre-construction inspection as well – this way any problems observed can be attributed to the construction work, and preexisting conditions can be excluded.

Just as PACP exists for pipelines and MACP for manholes, the Lateral Assessment Certification Program (LACP) was created by NASSCO to standardize defect coding for lateral inspections. Codes for defects in LACP inspections largely mirror those used in PACP inspections, easing the cognitive burden on operators. Further, software packages often integrate PACP and LACP inspections, making it easy to start a new LACP inspection when a lateral launcher is sent out during the course of a mainline inspection. Location along the host pipe, nomenclature, date, and other information can be automatically populated. This enables LACP surveys to be easily attached to house numbers or street addresses for follow-up with homeowners.

LACP can be slightly more cumbersome than PACP, as it relies on the creations of new inspections for each pipe segment. This means that every time a wye or split is encountered, the existing survey must be ended and a new survey begun. Thus, if a single property has a wye connecting an abandoned roof drain connection and a sewer connection from the house, three unique LACP surveys would be generated: the first would run from the public sewer line to the wye, the second would run from the wye to the cleanout or termination of the abandoned roof connection, and the third would run from the wye to the cleanout for the connection to the interior of the house. While this maintains a sense of organization, inspectors typically engage with one segment of pipe at a time, making it easy to lose sight of the overall context. Perhaps software vendors will take it upon themselves to incorporate LACP functionality into a system that can create a network overview of a PSL – giving homeowners easily digestible reports and making sure operators don't forget to inspect some branch of a service line.

While the need for lateral inspections is significant, the fact that they are privately owned can produce challenges for municipalities looking to take on some of the responsibilities of maintenance. First, accessing private property without the permission of a homeowner can present serious legal issues, and all PSL inspection programs should involve formal notice to homeowners and be reviewed by a solicitor prior to commencing. While the case can be made that investigating these laterals is in the public good, damage can and does occur during an inspection. What if a launched lateral camera gets stuck during the inspection? What if the location is under a poured concrete driveway? Remedying this situation will require a significant amount of repair and remediation, and will only be exacerbated by a homeowner who was not made aware of the work.

Further, if inspection indicates the presence of issues, the question of who should pay for repairs has become significant. Many property owners would be willing to pay for repairs, especially if presented to them in a friendly manner. Voluntary compliance can be especially high if costs can be financed and paid for in installments on a monthly service bill. Enforced compliance can be difficult, as elected officials may be unwilling to take on the loss of political capital that could come from requiring certain residents to pay for these repairs. Some sewer authorities have spent considerable amounts of time and effort to educate elected leaders in the hopes of obtaining regulatory support for requiring property owners to maintain and repair their laterals. Sometimes these efforts succeed. Sometimes they do not.

In other cases, municipalities are willing to take on the cost and burden of repairing PSLs in the guise of reducing costs for all ratepayers. However, even with willing public officials, repairing these lines can present legal and logistical challenges. Municipalities would be spending public money on privately owned assets, and if work is not distributed evenly across a service area, questions of bias and preferential treatment can arise. Allowing municipalities to use the scale of government contracts can reduce the cost of repairing these service lines, but homeowners may not be happy with the level of service, especially if surface disruption is required.

A happy medium seems to exist in the area of warranty or insurance programs. In many areas, homeowners retain ownership of their PSL, but pay a small warranty fee as a portion of their monthly service bill. This warranty fee builds up a reserve fund at the water authority that can be used to repair PSLs for any reason – blockages, I/I, etc. If the insurance can be applied to all property owners, questions of fairness and preferential treatment can be eliminated, and ratepayers funding the system can help allay costs of using public resources to help improve individual properties.

OTHER STRUCTURES

Many technologies discussed in this text can trace their origins to devices developed for pipelines – CCTV crawlers became lateral launchers; lasers used in pipes are now used to model manholes, etc. However, this modification is typically only suitable for other quasi-linear structures: in order to inspect a pipeline, a robot needs to move in one dimension, along the axis of the pipe. In order to assess the condition of a manhole, a device still need only move in one direction – the axis of the manhole. Devices need only be modified to move vertically, rather than horizontally.

There is a category of structures for which these simple modifications do not apply. Large structures – such as storage tanks, wet wells, and even some undersea outfalls – cannot simply be scanned through linear motion. Instead, devices must scan the structure in two or even three dimensions in order to assess its overall condition. The entire interior of a storage tank or wet well must be inspected in order to detect potential structural issues. An undersea outfall must be examined externally to check for risks or early stages of damage. Crawlers, push cameras, and the technologies that have been discussed in previous sections are ill suited to these tasks.

Robotic crawlers are designed to move through pipes in a linear fashion. In some devices, navigating turns and bends requires cumbersome skid steering, and the precise movements needed to accurately map a two-dimensional space can be extraordinarily difficult. Further, crawlers kick up silt and debris as they move through a space, making visual inspections of fully charged structures an exercise in patience that does not lend itself to thoroughness. Push cameras, SmartBalls, and passive floating platforms are of no use in these settings either, as there is no way to move these units in the required directions.

In many cases, the job of inspecting these structures falls to highly trained divers. These individuals are equipped with self-contained underwater breathing apparatuses (SCUBA); wet or dry suits; and man-portable inspection equipment, including eddy current sensors, high-resolution cameras, and more. Divers slowly make their way through these structures, assessing condition (and sometimes repairing flaws with underwater welding equipment) as they work their way through a unit. In other cases, structures will be dewatered in order to facilitate larger manned inspection crews.

This work is extraordinarily dangerous – even in dewatered conditions. When tanks are drained, toxic gases can build up inside, and the change in pressure and moisture can lead to the rapid propagation of failures. Working in partially or completely dewatered wet wells can require hoists, scaffolding, and harnesses, and a plug failure can cause circumstances to change rapidly. Sewer workers and trained divers have drowned in these situations. While entry may be unavoidable for maintenance, entry for the sole purpose of inspection should be avoided if possible. In any case, proper confined space procedures should be followed every time, without exception.

While divers may not have to worry about breathing in toxic gas, *in situ* inspection is no less dangerous. Even well-trained divers can easily become disoriented in confined spaces and are always working against the clock, as air supplies are finite. Bulky equipment can get stuck on materials and agitation or panic can increase the rate of oxygen consumption. Between 2008 and 2013, 54 divers died in the process of inspecting tanks and other structures. All of these divers had some kind of specialized training, beyond their SCUBA certification. Many deaths occur when would-be rescuers, eager to assist, thrust themselves into the same hazardous situation that imperiled the first victim.

Even worse, in developing countries, divers are often the primary technology used to inspect wastewater systems. In these countries, divers often enter manholes and pipelines with no protective gear, and hold their breath as they swim through the system, dislodging clogs, removing debris, and performing other maintenance activities that would be unthinkable in the United States.

While crawlers and robots have largely replaced sewer divers in pipelines in countries that can afford the technology, there is still a strong need for inspection technologies that have the flexibility of a human diver, but can be operated remotely, from a safe location for use in large tanks and chambers. The dominant technology in this space involves the use of remotely operated vehicles (ROV), which are teleoperated undersea devices that can be controlled by an operator at the surface. ROVs were first developed by the U.S. Navy for undersea exploration, and gained increasing popularity in the oil and gas industry. ROVs typically have neutral buoyancy and a series of propellers that enable them to be steered in multiple dimensions.

ROVs are well suited to inspecting the exterior of offshore oil platforms and large undersea pipelines. Because they are linked to the surface with cables, they are not limited to onboard battery power and can endure missions longer than typical divers. They also can be constructed from materials rated for extraordinary depths – this combination of material strength and agility enabled the ROV Jason Jr. to explore the interior of the Titanic. Jason Jr. was able to navigate inside the tight passages of the ship at a depth of more than 2 miles below the surface.

While ROVs are well suited to these exploratory tasks, they do have some shortcomings. While these propellers enable agility, most are not rated to fight intense flows. Thus, ROVs are suitable for inspecting large chambers with stationary or minimal currents, or the exterior of surfaces in larger bodies of water – not pipelines with significant flows. Further, cable management can be significant, as movement through water in three dimensions can cause the cable to be twisted to the point where internal shorts and breaks develop. This is more of an issue in open-water deployments, but can occur in confined spaces, as well.

ROVs primarily rely on CCTV for vision, so their utility in wastewater spaces can be challenging. The high turbidity of many flows can pose a major hindrance to navigation, though many contractors have used the devices in the southern United States to explore stormwater structures. To compensate for this lack of vision, ROVs can be equipped with state-of-the-art multibeam sonar for detailed mapping, though the low refresh rate of sonar may make it difficult to rely on for real-time navigation. ROVs can also come furnished with manipulators that can enable them to move objects, grab onto surfaces, or take samples of debris.

While the ROVs used to explore the Titanic remain expensive, advances in miniaturization of components have reduced the overall cost of standalone units considerably. Small ROVs are available from Deep Trekker for less than $5,000, while more advanced units can be had for less than $30,000. Utilities wishing to use an ROV as part of a service can turn to contractors like Hibbard Inshore, which will deploy the devices as part of a preplanned mission for either a quoted price or daily rate. While the price of an ROV inspection service may tempt municipalities to purchase a system and attempt to do the legwork themselves, remember that service contracts can contain performance clauses. If a contractor doesn't collect the data that they say they'll collect, they don't get paid. If staff can't figure out how to use the ROV successfully, costs can be difficult to recoup.

10 Software Packages and Asset Management

The previous chapters have discussed the different types of tools and methods that could be used in condition assessment programs for various types of buried infrastructure. Each and every one of those tools generates a significant amount of data, the management of which is key to making effective prioritization decisions and operational plans. However, there are a wide array of software packages available that can utilize this data – some ranging from large-scale, enterprise-grade packages to smaller tools that are licensed for only a single desktop. Some tools are intended to interface with multiple departments and aspects of the municipal government, while others are intended for their functionality to stand alone.

All fall under the broader category of asset management software packages. While some consider asset management to be the domain of financial experts and accountants, the everyday tasks of those responsible for buried assets – such as repair, replacement, and rehabilitation decisions – fall squarely under this umbrella. While many agencies are used to reactively responding to failures, blockages, and other disruptions, incorporating a more proactive strategy can save time and money. These software packages help water and wastewater professionals make sense of condition assessment data and work toward a systematic approach to asset management. Understanding the capabilities and limitations of these software packages can help prevent IT and data nightmares in the future.

This section will discuss CCTV-oriented inspection platforms, financial decision support packages, geographic information systems (GIS), and enterprise asset management (EAM) platforms. This discussion will focus on ways in which these broad categories of tools can interact with one another, though relevant vendor-feature functionality will be highlighted where appropriate. This is not a substitute for a software user's manual, instruction guide, or paid training. Rather, it will focus on describing categories of software to assist in deciding which type of system may be appropriate for the needs of a particular municipality or water authority.

ASSET MANAGEMENT OVERVIEW

Before describing the nuances of individual software packages, some discussion of the principles of asset management assist in understanding the context that led to their creation. Many organizations define asset management differently to meet their bespoke needs, but most definitions frame the discussion around finding answers to seven key questions outlined in the International Infrastructure Management Manual. Municipalities with an asset management plan should be able to confidently answer:

1. What assets does the municipality have?

2. What are the assets worth?
3. What is the condition of the assets?
4. What needs to be done to the assets in order to maintain an acceptable level of service?
5. When do these actions need to be taken?
6. How much will this activity cost?
7. Will this activity be financed, and if so, how?

Effectively answering all of these questions means that a municipality not only has an accurate picture of what exists in the system, but also a plan for what needs to be done to continue effectively providing service. While the unexpected can and does occur, taking a statistical view of a system makes it possible to plan for failures and respond accordingly, rather than be blindsided at every occurrence. The thought of moving from a reactive model to a comprehensive asset management program can be daunting, but effectively managing condition assessment data can help answer several of these questions – a task that is made much easier with decision support software.

Surprisingly, many municipalities find that they need to start with the first question: what do we have? While it seems like the answer to this question should be obvious, everyday CCTV surveys result in found manholes, and new construction projects hit lines that were not marked on any maps. Manhole inspection programs that begin with GPS location are commonly used to rectify this, though much will still be discovered in the course of pipeline inspections and other operational activity.

Even if a municipality knows to a fair degree of certainty, where should that information be kept? An Excel file on the public works director's desktop is probably not the best choice, even if it can be easily constructed. When manholes are found, or new construction occurs, how can this data set be updated? What kind of information is stored – is it a list of manhole ID numbers? Construction data? Can surveys be linked to this information? When one considers analyzing this information to get statistics like the average asset age, date of last inspection, or defect count, it becomes clear that more is needed than an Excel sheet.

Condition assessment can also play a huge role in answering the third, fourth, and fifth questions. Inspection programs return data that literally indicates the condition of the assets, but it often takes a trained engineer to answer questions about what needs to be repaired, how those repairs should take place, and when individual repairs should be scheduled. The PACP quick scores provide a useful rule of thumb for analyzing the condition of buried infrastructure, but they should not replace engineering judgment. While nobody will dispute the severity of a Level 5 defect, a Level 4 defect under a major highway may take priority over a Level 5 defect in an isolated woodland area.

The right software package can make it possible to integrate different sources of condition assessment data in order to answer these questions more thoroughly. For example, a pipeline may have only Level 3 defects, but a laser inspection may reveal a significant amount of wall loss. Tasking an engineer with analyzing a CCTV database and a separate laser database can complicate a task as he or she can get bogged down in the nuances of file structures, making it difficult to focus on constructing a lifespan estimate. The right tool will present both sources of data to the person making the analysis at the right time. This concept is termed decision support. The best software packages aren't the packages that show the most information or have the lowest price tag – they are the packages that are designed to work in a way that helps users. They contain features that can toggle information on or off and group it logically to help answer questions as they are being asked.

For example, a site inspector reviewing pipelines for new construction acceptance should be able to see all lines that meet certain criteria (constructed in the last month) in a single display, and not have to locate each one individually. He should also be able to see at a glance which lines passed acceptance testing and which lines may need a closer look. Drilling down into each of those lines should make it possible to view the CCTV inspection, as well as any laser data (such as a virtual mandrel test) corresponding to that line. It should be easy to jump to problematic spots or areas to determine the issue – did the laser indicate ovality because some debris washed into the pipe, or does it look like deflection is truly outside acceptable parameters?

Technically, software is not required to perform any of these analyses as these tasks can be completed by hand, but the right software can dramatically reduce the time it takes to answer these questions. Some may argue that engineers should earn their bill rates by manually sorting through this data, but remember the number of hours spent on a project can quickly exceed the license cost of a useful tool. If an engineer is on staff, remember too that there are only so many hours in a day. If he or she doesn't finish an analysis in time, pipes could fail, service could be disrupted, and individuals could be injured or even die. Asset management is not the place for a turf war – word spreads quickly about useful tools. Aim to be a center of competence.

Questions 2, 6, and 7 often require the involvement of financial analysts, but their success will depend on having accurate condition assessment data. Standard depreciation models may be spread over a 30-year lifespan of an asset. If condition assessment data shows that corrosion is slower than expected (or faster), this information needs to be made available to the financial team so that they can adjust their models accordingly. Similarly, developing accurate rehabilitation quotes or estimates hinges on the availability of accurate condition assessment data – the presence of which can reduce costs significantly. If a laser inspection reveals that the actual interior diameter of a 60″ pipeline measures to 62″, a liner can be designed to maximize the use of annular space. Further, condition assessment data makes it possible to model several different approaches – say 52″, 54″, and 58″ liners – to maximize available capacity and minimize installation headache.

While day-to-day personnel may be somewhat insulated from discussions of rates and bond issues, no department likes to be called out as a center of waste in a city council meeting. Asset management software can help answer questions in a way that saves time and money – sometimes defying conventional wisdom. Small savings in O&M budgets can translate to large gains in capital budgets, as removing small recurring costs can free up significant amounts for strategic programs. This multiplier effect will be different for each municipality, but don't be surprised if an operational savings of a few hundred thousand dollars can enable millions of dollars in capital projects without raising rates.

This link between condition assessment and financial planning is a major reason why software packages exist at different scales. CCTV-centric tools may be useful for planning on-the-ground operations within a water and wastewater department, while city management or budget directors may want to develop a picture of operational spending that goes beyond pipelines – to include roads, bridges, buildings, and other assets. Thus, finding software packages that provide decision support utility and link to or connect with other tools is a critical component of enabling intergovernmental collaboration.

Effective asset management can be hindered and helped by a variety of forces. The EPA's National Pollution Discharge Elimination System (NPDES) has done much to institutionalize the response to combined and sanitary sewer overflows. Consent orders have made wastewater authorities formalize their approaches to wet weather planning, maximizing available capacity – in order to serve ratepayers and demonstrate compliance to the federal government. The Government Accounting Standards Board also published a rule – Statement 34 (GASB 34) – that provided new standards for municipal accounting regarding assets that could be classified as physical infrastructure. Among other things, GASB 34 required municipalities to maintain an accurate list of assets, assess their conditions every three years, and estimate spending to maintain a given service level. Complying with this regulation requires a well-structured condition assessment program, and tools to manage the data it generates. The combination of NPDES and GASB 34 has led to many municipalities adopting comprehensive condition assessment programs on a recurring basis.

Further, many municipalities cited two key factors inhibiting progress when surveyed by the Government Accountability Office about their asset management programs. The number one factor was limited budgets or a lack of financial resources. Unfortunately, infrastructure has been neglected on a nationwide basis – years of deferred maintenance have led to the current state of decay. Creating an asset management program can sometimes cause sticker shock: authorities generally know the current state of a system, but it can be surprising to see how that translates into dollars – the cost required to restore or maintain a certain level of service. Seeing this financial reality didn't suddenly create the need – it only quantified something that many may have been vaguely aware of. It also highlights just how acute these shortcomings may be when compared to standard annual budgets.

The other factor – which is closely intertwined with limited budgets – is the short-term planning horizon of elected officials. Politicians are creatures that respond to the same incentives that apply to the rest of humanity. They do things that bring them rewards, and they avoid things that incur costs. Unfortunately, buried infrastructure is not a popular topic for politicians. While infrastructure spending is a topic that engenders rare bipartisan support, most elected officials would rather be attending the ribbon-cutting ceremony for a new bridge than explaining to local businesses why a major interceptor needs to be replaced. Unfortunately, this causes many to simply kick the can down the road. While the interceptor needs to be replaced, it probably won't fail in the next year, and a councilperson may be safely reelected by that time and can reconsider the issue then. Unfortunately, when that time comes, the official decides to double down and defer maintenance again and again until failure ultimately occurs.

While it is obvious that sinkholes can end political careers, asset management programs provide ammunition in the fight for political support for water and wastewater infrastructure. Confronting an elected official with the disparity in actual spending vs. required spending, along with a summary of consequences that could occur, can be a powerful tool. Every department needs something, and turning down "badly needed" interceptor rehabilitation funding may mean enabling "badly needed" road projects. Turning down funding for rehabilitating an interceptor that has an 80% chance of failing in the next year, costing the municipality more than $50 million in emergency repair costs, is much more difficult, as no elected official wants to carry the burden of knowingly increasing the risk to constituents.

It is poor form to simply make up frightening sounding projections. Fortunately, asset management software tools can help you create these statistics and cost estimates based on real-world data. This can begin at the level of the individual CCTV inspection.

CCTV MANAGEMENT PLATFORMS

One of the earliest types of asset management tools to emerge were programs that specialized in managing CCTV inspections of pipelines. While NASSCO's PACP standard was designed to enable paper-based completion of condition assessment projects, computers have revolutionized the accuracy of CCTV surveys. Software packages can prevent users from applying incorrect modifiers to codes, ensure that continuous defects are closed appropriately, and automatically link records with relevant images and timestamps in a digital video file. This can save hours of time for field crews, office staff, municipal customers, and anyone interacting with files in the data stream (Figure 10.1).

These software packages were initially designed to manage the operations of a CCTV inspection truck, enabling users to perform multiple surveys, label data files appropriately, and export collection to an external drive or cloud server for delivery to a client. As these tools became more widely used, vendors began developing specialized office versions of the software that could manage large sets of inspection data. These office versions started out as simple viewers, but have gradually evolved into fully featured applications that can manage inspections from dozens of crews simultaneously. Many integrate with GIS or EAM software, and have become a staple of public works departments, engineering firms, and water and wastewater management teams.

Evolving the feature set of office applications is a shrewd move on the part of software developers as it can lead to a particular tool being included in the specification for CCTV inspection work. This may come as a surprise, as PACP defines an exchange format, so that databases generated by one platform can be automatically imported into another. Therefore, if a contractor is using WinCan for collection and the office staff is using PipeLogix for review, data can be freely exchanged between the tools. While this is technically possible and may be essential for merging legacy data sets, in day-to-day operation, importing and exporting CCTV surveys between different software packages is untenable for a generalist. The PACP 7.0 standard

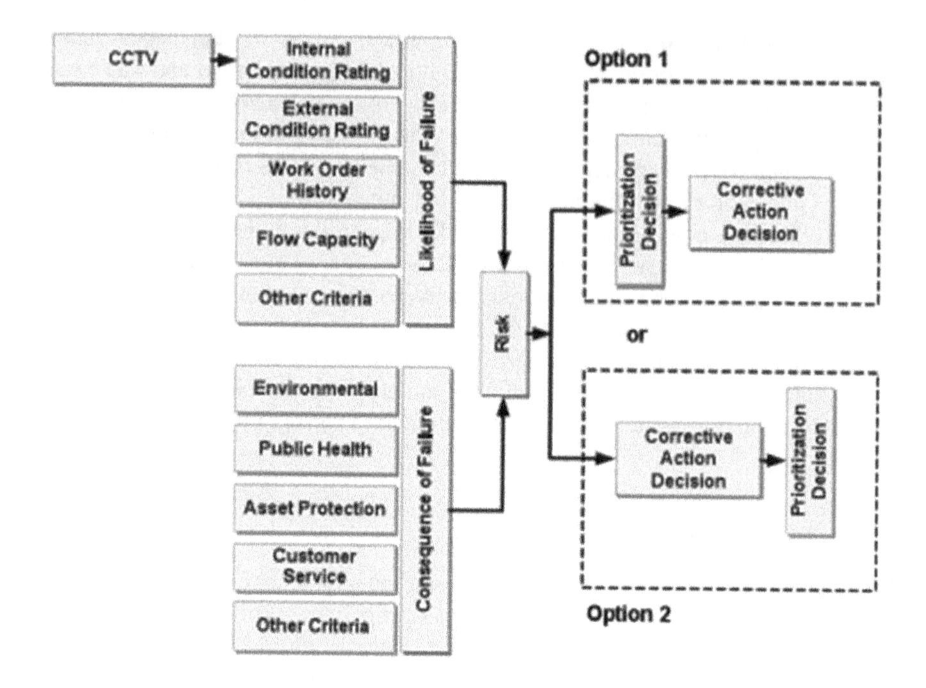

FIGURE 10.1 CCTV-based systems use inspections as one of the primary tools to determine the likelihood of failure of an asset. Many of these platforms derive from software that is already used on inspection trucks such as WinCan, Granite XP, or ITpipes. Public Domain: https://nepis.epa.gov/Exe/ZyPDF.cgi/P1008CD6.PDF?Dockey=P1008CD6.PDF

defines required features of an inspection, but each vendor implements major quality of life features in a slightly different way. Two or three extra clicks to make a survey appear properly may not seem like a big deal, but can add up to hours of lost time when repeated across dozens or hundreds of inspections – or worse, errors in analysis if anyone ever forgets to take the necessary manual steps. Perhaps if a database expert or dedicated IT staffer were available to handle these data conversions, any compatible software package could be used, but why bother when using tools from a single vendor makes data aggregation and analysis so simple?

Due to their origins as truck-based platforms, software packages like WinCan, IT Pipes, Granite, and PipeLogix have historically excelled at managing data at the inspection level, and many are designed to enable efficient collection of survey data by individual crews. Many tools will auto-complete complex manhole identifiers, prompt users to enter the required clock positions and modifiers for codes, and automatically start a new survey at the end of the previous one (taking the ending manhole of one and using it as the beginning manhole of the next). While these features may not sound significant, they can dramatically reduce errors in data and improve crew throughput. Further, some errors have a permanence that can be difficult to correct. If a manhole name is typed incorrectly, it can end up permanently encoded in a video file, even if it is later corrected in the database. If a video is viewed out of context years later, this could contribute to confusion or even an incorrect rehabilitation decision.

From the perspective of the CCTV truck operator, there is little choice in choosing a software package. Initial inspection management software tends to come with the truck. Cues trucks will be outfitted with Granite, RedZone Solo robots are integrated into ICOM, and Envirosight devices are often bundled with WinCan. The software that came with a truck is usually closely tied to the outfitted hardware, so support issues will be minimal. For private, emergency, or other one-off inspection work, this software will likely be sufficient. All major packages allow for the production of PDF reports and export inspection video truck, images, and databases to a USB thumb drive for delivery to a client.

Contractors would do well to explore other options. As noted previously, many municipalities specify a particular CCTV inspection software in their televising contracts. Being able to equip an inspection truck to switch between WinCan, Granite, POSM, and other platforms will make one eligible to bid on and participate in more inspection work. Many of these software packages install specific drivers for video capture hardware, screen writers, and USB controllers; so setting up multiple systems is usually not as easy as installing the programs side by side. Instead, consider outfitting the truck computer with multiple hard drives – each containing an installation of a different CCTV inspection software package. This will reduce conflicts and make it easy to simply choose a new tool by restarting the computer.

Some contractors try to get around this by hiring a team of database experts to work in the office and convert data between formats. While having some database expertise in-house can certainly be a value-add, relying on an individual to manually facilitate data conversion will 1) encourage data entry or data interoperability sloppiness (especially in table names and IDs) 2) slow down delivery time by introducing an additional step in the process (making off-truck delivery impossible), and 3) increase costs over time, as many software packages can be updated for a one-time fixed cost, whereas the cost of a fully-loaded staff member will continue for years to come. From the municipal specification perspective, this behavior can be discouraged by requiring collection to be done in a specific tool of choice, without exception. Any contractor eager for the work will find a way to obtain the software needed to complete a job properly.

An additional note about the computers on CCTV trucks is in order. Many modern platforms enable data to be synchronized with the cloud and recommend an Internet connection. With this connection to the outside world comes the need to follow IT best practices – make sure that all CCTV truck computers are up to date with Windows updates and software patches. Consider running an antivirus program that routinely scans for infection (Malwarebytes, F-Secure, and Smadav are good tools that require minimal CPU/RAM in terms of overhead). Moving data onto or off the computer will require the use of USB hard drives or flash drives, so the threat of viruses is always present. Crews with idle time will naturally browse the Internet and check email while waiting for a robot to be cleaned or files to copy – making it a matter of time before a malware infection results. Consider removing Internet browsers from the truck computer to eliminate this possibility.

From the municipal or office perspective, the decision can be a bit more complicated. Many municipalities begin by using the office version of the software that comes with CCTV inspection trucks. If all trucks come with WinCan, then WinCan ends up being the tool used for data review and analysis. However, it can be useful to regularly evaluate this decision. The capabilities of software packages change dramatically from version to version. The evolution of WinCan from version 7 to version 8 and VX completely changed the look and feel of the program and added incredible amounts of functionality. Similarly, ITpipes Web is a vast improvement over the older desktop edition (version 2) of the software. Regularly invite vendors to demonstrate their software and always get updated pricing – sometimes the cost to switch to a new platform can be less than the cost of a routine upgrade.

Any table or matrix comparing features available in CCTV inspection management software is certain to quickly go out of date. Therefore, rather than compare features as they exist at a point in time, use these questions to guide an adoption decision. First, while CCTV-centric software is useful for inspection management, how well does it integrate with other tools, especially EAM systems? If a city uses IBM Maximo or Cityworks or Hansen, how easily can data be exchanged between the EAM system and the inspection tool? Almost all providers will claim interoperability, but many rely on what some consider a technicality to justify this. By manually exporting a CCTV inspection and importing it into the EAM, some form of data exchange will always be possible, but useful interoperability will rely on seamless data exchange using a software tool or module.

Is it possible to upload data from the office edition of a CCTV inspection platform to a city-wide EAM? How many clicks does it take to do so? Can data be downloaded to the CCTV software, so that work orders can automatically queue inspections? Can the vendor demonstrate this functionality today, or would they develop it as part of the contract? Has the vendor provided this functionality for a specified EAM in the past? Can that customer speak to the experience? It is worth digging deep into the answer to this question, as data exchange and interoperability will be a daily activity for water and wastewater departments in most medium to large municipal governments.

Similarly, what level of GIS integration is provided? Sending inspection crews out into the field with maps can make inspection work much easier, and being able to view inspection data in the city-wide GIS can make planning repairs much easier. Not all tools approach GIS integration in the same way. ITpipes integrates with ESRI's ArcGIS by default – a great time-saver if a municipal government is already using ArcGIS. By contrast, RedZone's ICOM comes bundled with its own GIS database – significant feature for municipalities that have not yet made the leap to GIS, but a potential complication for municipalities that may already have extensive geodatabases that are regularly updated. Creating the snapshot required for RedZone's ICOM may require a significant amount of synchronization work at a later date to merge back into the master database.

Finally, walk through the workflow of reviewing CCTV inspection data. What does it look like to aggregate multiple inspections from multiple trucks? Will an inspector be responsible for reviewing a stack of hard drives, or can the software provide a glimpse of what inspections have been performed through the synchronization of a cloud database? If the data is accessed via cloud, what is the performance like? Can a user scroll through video without buffering? If the data is accessed locally, how can it be shared among multiple users? Are file servers supported? Are databases locked upon access or can multiple users load at the same time? Is there a cloud version of the software in the roadmap, and if so, at what cost?

A demonstration version or trial of the software can immediately highlight pain points that would never be covered in a sales presentation. Many CCTV-centric software packages enable users to name manholes on the fly – thus while start and end manholes may correspond to their GIS names, any found manhole may be named at the discretion of the crew. Thus, a disproportionate number of these newly discovered manholes are named MH 1 or Manhole A. How a software package treats these potential duplicate entries can be critical in terms of maintaining data integrity and assigning inspection records to the right real-world assets.

When negotiating the adoption of a new software package, consider the cost of importing legacy data in addition to software license fees. Many vendors will be willing to throw in importing legacy data with the purchase of a software subscription – but not unless requested. This can be a great way to quickly get all data onto a single platform – since software vendors are intimately familiar with databases and data structures, they can often complete a data import project much faster than a typical team of interns or junior engineers. Most of these efforts will be limited to digital files – VHS tapes and paper records often require an added professional service fee.

Many municipalities will request that CCTV software vendors support a custom set of codes that goes beyond PACP in some way. Sometimes, this interest is well founded – for example, the use of WRc in Canada or WSA in Austrialia. However, other times, the custom code set may exist for historical reasons that are no longer relevant. Rather than continue using a particular piece of software due to its support for nonstandard defect abbreviations, consider investing in a one-time data mapping project or think of other ways to reframe an analysis that does not depend on custom code sets. Custom codes will limit the pool of available contractors for inspection work, can promote data quality issues, and will make it difficult to integrate data from other municipalities into a comprehensive analysis.

In the management consulting space, companies are often advised against reinventing the enterprise. If their business doesn't fit into the functionality provided by Salesforce.com or Netsuite, they should consider reorienting their business, rather than creating a new, incompatible CRM. The same holds true for PACP codes. Municipalities should consider revising their analysis process to be compatible with PACP codes, rather than continue to rely on an obsolete and nonstandardized code set. PACP is not perfect, but it is well understood by consulting engineers and contractors. Municipalities adopting it buy themselves flexibility in terms of new hires, consultants, and external resources.

ASSET MANAGEMENT PLANNING TOOLS

For a small municipality, the office edition of a CCTV inspection management tool may be sufficient for aggregating assessment data, issuing repair orders, and even planning and scheduling work for weeks or months in the future. However, these software packages do not include tools for budgeting and financing these repairs from a municipal perspective. Large-scale EAM packages can bridge this gap, but are not accessible to smaller municipalities due to the high price of software licenses (though this is improving). Fortunately, several free tools have been released by federal and state governments, specifically targeting the needs of small governments and water authorities.

Some may argue that specialized software is not required for creating a basic asset management plan – a simple spreadsheet with the right formulas could do the job. Several state governments agree with this assessment, and have released asset management templates that can be loaded into Microsoft Excel – users need only populate asset types, costs, and estimated lifespans in order to generate a useable plan. These spreadsheets can be downloaded free of charge and can be modified as municipalities outgrow the built-in functionality.

Pennsylvania's Department of Environmental Protection (PA DEP) has released a workbook that is typical of this class of tools. Their "AM Basic" spreadsheet uses a combination of prebuilt formulas and macros to automatically generate a fiscal forecast by estimating cash needs and revenues based on entered parameters. Users can enter up to 200 individual assets, along with the year of installation (if known), replacement cost, and a life expectancy estimate. If life expectancies are not known, typical estimates for computers, force mains, motors, pumps, hydrants, and other fixtures are included (Figure 10.4).

This information is used to generate a list of cash needs for existing infrastructure. Users can add additional costs in the form of O&M expenses, debt service, or other items, and balance those costs against forecast revenue from ratepayers, municipal budgets, or grant funding in order to estimate an annual surplus or deficit. A 20-year estimate of funding needs is automatically generated by the spreadsheet – suitable for presentation in public meetings. Municipalities may find the spreadsheet slightly limited, but can copy and paste the sheets into a new document, and modify the formulas and macros accordingly. If the sheet does not appear to be operating, make sure macros are enabled in Excel – otherwise charts will appear frozen and will not update automatically.

With tools like the Pennsylvania AM Basic spreadsheet, don't let the perfect be the enemy of the good – enter as much information as is known, and return to refine and improve in the future. These tools are designed to get an asset management plan off the ground – not to create a perfect, nuanced budgetary estimate in an afternoon. Some asset management plan is better than no asset management plan, and a spreadsheet like this can provide an excellent starting point. Also note that the PA DEP tool does not contain any information specific to Pennsylvania laws or regulations – it is suitable for municipalities or water systems in any state looking to get started with asset management planning.

While not necessarily a tool, it is worth noting that resources exist to help small water systems create and implement asset management plans – at no cost to the system. The Environmental Finance Center Network (EFCN) will provide free consulting to small water systems (defined as those serving populations under 10,000) on a variety of financial topics. While this includes asset management planning, it also encompasses rate setting, long-term capital planning, minimizing nonrevenue water, and other strategic initiatives. In addition to one-on-one consulting, a variety of free webinars and training courses are also available. The EFCN is a collaboration sponsored by 10 university centers and serves small systems around the United States. Other state-specific programs may also exist – check with your state's department of environmental protection or quality for more information.

For municipalities that may require more than a simple spreadsheet, the EPA offers a free software package that can perform more involved asset management planning. The Check Up Program for Small Systems (CUPSS) is designed to help smaller utilities develop more comprehensive asset management programs than the spreadsheets described earlier, which estimate spending based on averaging and replacement cost estimates – not the actual costs of emergency maintenance that are required to keep assets in service (Figure 10.2). CUPSS enables municipalities to target a desired level of service and calculates the cost required to do so.

CUPSS is often offered as part of a training program and is bundled with a workbook and user manual. These resources walk municipal managers through a process of goal setting, entering an asset inventory, logging service records, and inputting additional revenues and expenses (Figure 10.3). It breaks tasks down into small chunks, so that municipalities can chip away at creating an asset management plan with small-time investments – say 30 minutes per day, over a prolonged period of time. More importantly, CUPSS includes a variety of resources intended to spark conversations around asset management planning between staff members at a utility and decision makers who may be removed from day-to-day operations. Free training resources are also available online to walk users through getting started inputting data into CUPSS.

The University of Massachusetts Amherst has also released the CUPSS Mobile Assistant – an app for Android and iOS devices – that allows users to enter asset information on a mobile device, and export that information to a spreadsheet that can

Check Up Program for Small Systems

FIGURE 10.2 While dated, the Check Up Program for Small Systems (CUPSS) is a great free resource for smaller utilities to get started with asset management. Unfortunately, there can be some challenges in getting the software to operate properly on modern systems. Public Domain: https://nepis.epa.gov/Exe/ZyPDF.cgi/P100NRWB.PDF?Dockey=P100NRWB.PDF

Drinking Water

Water Supply Connections

Residential Facilities connected to potable water Goals No.
Commercial Facilities connected to potable water Goals No.
Industrial Facilities connected to potable water Goals No.

The Drinking Water Network

Asset Type	Unit	Description
Wells and Springs	Number	1
Pumping Equipment	Number	2
Disinfection Equipment	Number	1
Concrete & Metal Storage Tanks	Storage Capacity Days	0
Distribution / Collection Mains	LF	0
Buildings	Number	1
Lab / Monitoring Equipment	Number	1

General Water Supply Information

Number of Connections 33 No.
Storage Capacity MG
Reserve Storage Days
How Sourced Descr.
Interconnected Descr (if yes)
Water loss and Inflow / Infiltration Descr.
Total Volume Produced Gallons /Day
Total Volume Sold Gallons /Day
Average / Peak Daily Consumption Gallons /Day

Water Supply Asset Values

Replacement Value $431,450 $000,000
Depreciated Replacement Value $245,932 $000,000

Wastewater

Wastewater Connections

Residential Facilities connected to potable water Goals No.
Commercial Facilities connected to potable water Goals No.
Industrial Facilities connected to potable water Goals No.

The Wastewater Network

Transmission Mains	Number	1
Valves	Number	1
Transformers / Switchgears / Wiring	Number	1
Motor Controls / Drives	Number	1
Sensors	Number	1
Buildings	Number	1
Treatment Equipment	Number	2
Security Equipment	Number	2
Sewers	Number	2
Pressure Pipework	Number	1

General Wastewater Information

Treatment Plants No.
Treated Effluent Discharge Points No.
Type of Treatment Desc.
Discharge Volume Avg. Gallons/day
Interconnected Descr (if yes)
Water loss and Inflow / Infiltration Descr.

Wastewater Asset Values

Replacement Value $2,100,450 $000,000
Depreciated Replacement Value $1,255,787 $000,000

FIGURE 10.3 CUPSS walks users through the input of information required to structure a basic asset management plan. Estimates of lifespan can be used in place of detailed forecasting. Information can be added in small pieces to gradually assemble a program, without taking resources away from critical activities. Public Domain: https://nepis.epa.gov/Exe/ZyPDF.cgi/P100NRWB.PDF?Dockey=P100NRWB.PDF

be utilized by the CUPSS desktop application. This enables data about assets to be entered near the assets themselves – perhaps as an operator is going through a pump station or treatment plant, away from the main office.

While CUPSS is free and easy to use, the program does have some significant downsides. First, it remains tethered to the desktop. As a Windows executable, it is unlikely that CUPSS will ever be upgraded to a web-based or mobile application that can run on multiple devices. Second, the program is no longer supported by the EPA and is quickly becoming outdated. CUPSS used to have a vibrant trainer community and ecosystem supporting the product – most of these resources are no longer widely available. That said, the program is still available as a free download and still functions on Windows 7, 8, and 10, though it may require some configuration to run properly. Running the installer in compatibility mode (for Windows XP SP3) can solve many common issues. It is also possible to run CUPSS in a dedicated virtual machine, though the lack of support means its days are numbered. Still, for a municipality looking to get up to speed on asset management in a modular way, it may be worth considering.

Many EAM systems have made significant reductions in pricing to target the small customers previously served by CUPSS; however, no product has emerged to fill the same niche. Perhaps the EPA will open source CUPSS so that development can continue, but until that point arrives, small system operators can continue to utilize the existing executable so long as it is supported by Windows.

GEOGRAPHIC INFORMATION SYSTEMS

One of the most significant trends in municipal asset management is the adoption of geographic information systems by municipalities of all sizes. Where CCTV software examines a system through the lens of individual pipelines, and asset management tools focus on physical pieces of equipment, GIS databases attempt to place events and items in the context of the three-dimensional world. With GIS, everything is inherently tied to its location on a map, providing immediate spatial context and visualization. While GIS was once the domain of cartographers and mapmakers, it has advanced to the point where maps can become an integral part of many aspects of municipal government (Figure 10.5).

Broadly, a GIS is a system that represents different pieces of information in a way that allows for display in a mapping system. Its success and versatility is intrinsically tied to digital, computer-based presentation. Historically, maps have been a series of trade-offs, as they can only represent so much information. Streets, rivers, and major features may be represented in a two-dimensional coordinate system, and colors or shapes may be used to denote cities or political districts. At some point, it becomes impossible to show everything on a single map – a plot of government boundaries, mineral resources, watersheds, soil types, and population density would be too cumbersome to be useful.

As a result, municipalities often had to maintain vast archives of different maps – property surveys, water lines, road construction, mineral rights, crime reports, census data, and more – each in a separate document. Identifying connections between different maps was an exercise in manual labor, and often relied on luck as much as a trained eye. Hacks, such as sheets of transparencies, attempted to make it possible to coordinate information between multiple data sets, but if maps were produced at different scales or from different perspectives, alignment could be physically impossible.

Advances in computing power led to major research and government agencies exploring the intersection between computers and cartography. In the 1960s, Canada launched one of the first GIS systems to track land use applications throughout the

Enter dollar figures in the yellow cells. Gray cells are calculated cells and do not require entry. It is not necessary to enter negative values.

		2020	2021	2022	2023	2024	2025	2026	2027	2028	2029	203
Cash Needs	1. Rehabilitation / Replacement Costs from 'Asset Entry' spreadsheet	($0)	($0)	($0)	($0)	($0)	($0)	($0)	($0)	($0)	($0)	($
	2. System O&M costs											
	8. Misc yearly costs and/or profit assumed for private entity											
	11. Annual payment on debt											
	TOTAL CASH NEEDS											
Revenue	7. Estimated Revenue from ratepayers											
	9. Estimated Miscellaneous yearly revenue											
	10. Loan or Grant Proceeds assumed											
	TOTAL REVENUE											
	13. Surplus / Deficit (will be applied to reserve fund next year)											
	14. Reserve Fund Balance (Enter balance for start of 2020.)											

Note: This sheet is assuming that the asset replacement schedule and cost calculated from the 'Asset Entry' sheet will be applied as an expense in the recommended year of replacement. If are assuming a loan for the rehab/replacement for the calculated assets, enter that in the loan or Grant Proceeds Assumed row in the same year as the expense. Enter expected loan repayment costs in the "Annual Payment on Debt" row in subsequent years to account for that loan payoff.

If your system is a for profit entity, add the expected profit in Misc yearly costs row in addition to any other miscellaneous costs projected.

FIGURE 10.4 Some tools, such as Pennsylvania's AM Basic Spreadsheet, enable small systems to input assets, replacement cost, and estimated lifespan in order to generate a forecast of cash needs in the future. This particular tool is limited in terms of the number of assets that can fit in the sheet. Public Domain: https://www.dep.pa.gov/Business/Water/BureauSafeDrinkingWater/CapabilityEnhancement/Pages/AssetManagement.aspx

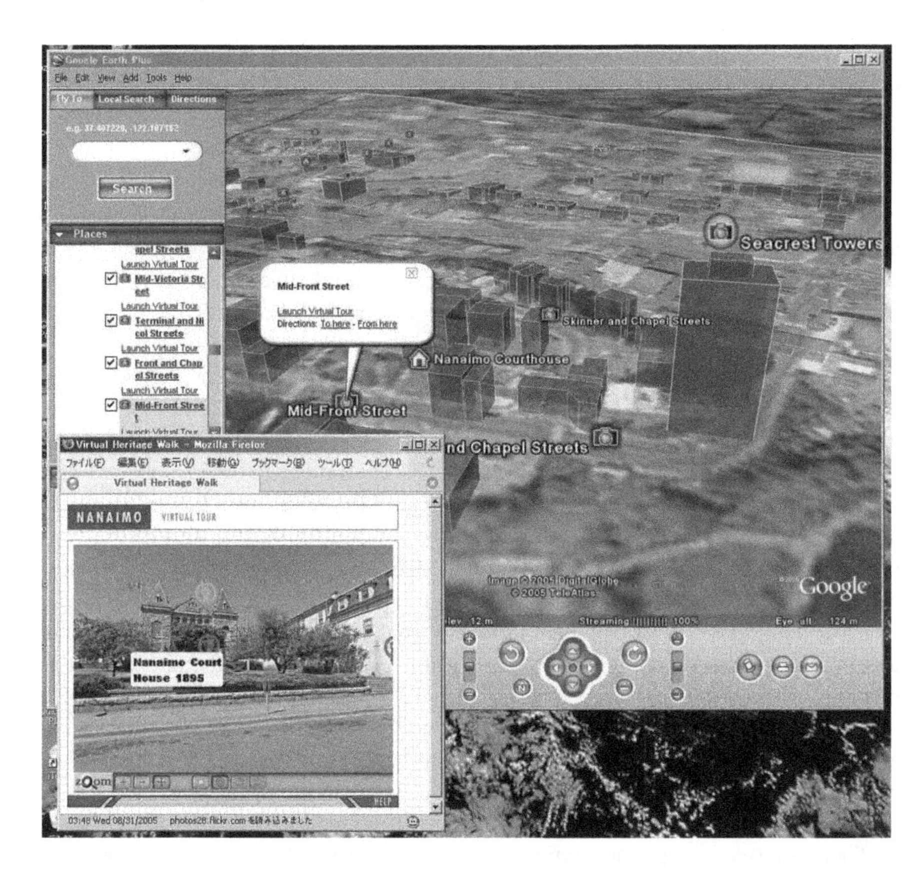

FIGURE 10.5 Geographic information systems provide a way to view water and wastewater data indexed to a map – so that anything displayed can be easily located. GIS makes it easy to integrate multiple mapping systems so that gas lines, water lines, and roads can all be shown on the same plot. This screenshot was created using Google Earth, but ArcGIS has become a municipal standard. QGIS is an open-source alternative. Screenshot provided by Earthhopper. Creative Commons: https://ccsearch.creativecommons.org/photos/7c7aeb3d-fba2-49b0-8fad-89e47ce56534

country. Running on mainframe hardware, the "Geo-information System of the Canada Land Inventory" was expanded over the subsequent decades to become a comprehensive resource. Its development was driven by a government geographer, Roger Tomlinson, who essentially invented the key principles that would go on to underlie the field of GIS.

Early systems like Tomlinson's were tied to mainframe computers and were not able to be broadly accessed by the general public. The Environmental Systems Research Institute (ESRI) began developing versions of GIS software that could run on smaller, less powerful computers – starting with minicomputers, and eventually moving to microcomputers in the 1990s. This desktop application, ArcView or

ArcGIS, has become so widely used as to become a de facto standard for engineering firms, municipalities, and federal and state agencies.

One of the key insights with GIS is the idea that data can be segmented into layers. A base layer may contain a graphical map, and layers containing other features may be added on top. A layer may contain zip code boundaries, water lines, work orders, or other constructs, and can be toggled on and off on demand. An engineer or user can construct a custom map by selecting layers in order to answer or investigate complex questions. For example, if one suspects a link between pipe corrosion and soil type, a map showing water main breaks and USGS soil classifications could be created to visualize connections between these variables.

GIS data typically exists in one of two formats: raster or vector. Raster data is analogous to a bitmap – an image that is constructed through the composition of multiple pixels. Pixels are arranged in a grid and each pixel can have a unique value – corresponding to colors or shades of gray. Modern digital photographs are constructed in a similar way – an image taken with a digital camera has millions and millions of these pixels, each with a unique color and brightness. In GIS systems, raster data is typically used as a background layer – say satellite imagery, for example, or even scanned maps or photographs.

Raster data has a level of realism that isn't present in vector data. Images made up of pixels can have a photographic quality and give the eye useful context for other pieces of information. When a raster image is the bottom layer in a GIS display, it is often termed the basemap. Rasters can also serve to represent data that varies along a surface – such as elevation plots, heat maps, or rainfall gauges. Individual pixels can change in value depending on this data source, and the cumulative effect of all of these pixels is to show variation across the entire landscape.

There are two major shortcomings to raster data: first, there is a resolution limit. Raster data exists in individual pixels, and if a map is analyzed at a level that goes beyond that unit, the result can appear blocky and difficult to understand. When one zooms in too closely on a digital photo, one sees an array of pixels, rather than microscopic levels of detail. When one zooms in too closely on a satellite image, large chunks of pixels appear instead of high fidelity detail of backyards and pavement surfaces. Some GIS systems can switch between different raster layers depending on the level of zoom, but this high end functionality is not a part of all systems.

The second major shortcoming is file size. Raster data takes up a tremendous amount of space – gigabytes upon gigabytes. While computers continue to improve, so do imagers, and high-resolution imagery is almost always the largest layer in a GIS. Processing this data in real time can lead to significant delays, so jumping a few blocks over on a map can require patience as the raster data is loaded from storage into memory. This problem is exacerbated by web viewers. While many municipalities have made GIS data available to residents through free web interfaces, these resources can be challenging to utilize, even with a high-speed Internet connection. FEMA's floodplain database is a good example – the system uses a high level of fidelity to provide accurate representations of different flood zones across streets and property lines. Moving from one parcel to another often requires hundreds of megabytes of data to be exchanged in memory for display.

Vector data is a much more common data source, as it is used to represent most municipal assets. In vector data, objects are represented as lines or points in space. These lines and points can be infinitely resized without losing fidelity and are ideal for representing roads, bridges, pipelines, and manholes. Even curved features like rivers can be represented by polylines – paths comprised of multiple smaller connected segments.

Vector data can be described mathematically, and as a result, takes up a very small amount of computational space. Points can be represented by a coordinate, and vectors can be represented by their start and end points. Files containing vector data tend to be a small fraction of the size of larger raster files. Because vectors exist in geometric space, vector data is often referred to as shape files or shapefiles (.shp). Shape files contain the mathematical information necessary to describe each shape, as well as any attributes or labels necessary for identification. For example, a segment of pipeline can be identified by the coordinates of the start and end manhole, as well as attributes for pipe size, material, year of construction, etc.

Vector data (and raster data to an extent) must be referenced to the three-dimensional world in some way. For example, sewer systems can be represented by lines, but those lines must be described in a way that ties them to the surrounding geography. In GIS, this is accomplished through the use of a coordinate system, also known as a reference system or datum. A datum is a reference system that describes the surface of the Earth in some standardized format – for example, the World Geodetic System 84 (WGS 84) defines the earth as an ellipsoid with a semi-major axis length of 6,378,137 m and a semi-minor axis length of 6,356,752.3 m. It uses the International Reference Meridian to define 0° longitude and the equator as 0° latitude.

WGS 84 is useful for describing global features and is used as a default geodetic system for many GPS devices. Other datums may be able to describe a portion of the earth more accurately than the WGS system. The North American Datum of 1983 (NAD 83) is frequently used as a local referencing system within the United States, Canada, and Mexico. While it generally agrees with WGS 84, points can be off by up to two meters between the two systems, making it critical for users to always check the datum of maps prior to integration.

Different coordinate systems also exist for states or portions of states to provide further accuracy. Florida, for example, has planes covering the northern, eastern, and western portions of the state. California is covered by 6 planes, while Alaska is divided into 10. These State Coordinate Plane Systems (SCPS) are assigned an identification number that GIS systems will refer to for consistency. Many of these spatial reference systems are stored in the European Petroleum Survey Group (EPSG) Registry where they are publicly available for analysis and comparison.

While the nuances of GIS can be challenging for a casual user to grasp, geographically oriented databases are quickly becoming systems of record for many municipalities. Updating the condition of buried assets and presenting that information on a system of maps can help establish an accurate view of what assets a municipality owns and exactly where those assets are located. Condition assessment data can then be quickly tied to a consequence of failure based on real-world evidence, rather than estimates and projections.

The link between manhole location programs and GIS should be somewhat obvious – by obtaining updated GPS coordinates of each manhole, structures can be accurately plotted on city maps. This can make it easy to dispatch crews in the future, maintain a single source of information for future underground work, and tie manholes to easement and property ownership information. For example, if a manhole is located in an individual's front yard, knowing exactly who to notify prior to preventative maintenance can save time and improve public relations.

Many surveys of buried assets generate information that can be used to update GIS information. For example, CCTV inspections will automatically note the length of the survey upon completion. This can be used to update the length of the sewer shape file in the GIS – something that can help locate a buried manhole or structure that is not visible from the surface. Remember though, tethers tend to stretch during deployment, and inspection length should not override the coordinates of start and end manholes. Both data sources should be in generally good agreement – differences of 3–5% are acceptable. Larger differences should trigger a QA check – was the encoder on the inspection device calibrated? Did it slip during the inspection? Was the correct segment actually inspected? This last question comes up quite frequently. If a segment is supposed to be 300′ long and an inspection is 500′ long – there is a good chance that something is amiss. Perhaps crews inspected the wrong pipe, or perhaps one of the manholes was incorrectly marked (maybe a telecommunications manhole was marked by mistake). Regardless, differences in inspection length of more than 5% provide a good opportunity to reestablish inspection procedures and double-check all data sources.

One issue that GIS systems will quickly make evident is the need for each structure to have its own identifier. From a computer science or database perspective, this need is obvious, as each record in a data structure must be able to be uniquely identified among all others (similar to a primary key in database terminology). However, many municipalities do not currently have such a scheme in place – instead, it is common for each interceptor or trunk line to have manholes 1–4. Part of creating a GIS system can be coming up with those unique structure names – perhaps the #1 manhole on the west interceptor could be named West_1, or maybe now is the time to come up with more descriptive names, which relate to chainage along the structure?

Pipelines are easier to deal with, as these structures are customarily broken down into segments that are defined by two manholes. Pipeline names can then be generated automatically according to a formula: for example, UpstreamManhole-DownstreamManhole. Thus, a segment running downstream from manhole West_1 to West_2 would be named West_1-West_2. Well-structured GIS configurations can handle this naming automatically. Some degree of forward thinking should be employed. Many regionalization programs have run into major logistical challenges when two systems merge, as manhole names were not as unique as once thought. Consider using names that have some basis in a geographic location to reduce the likelihood of inadvertent duplication, and be ready to compromise when working with neighboring systems.

While almost all CCTV inspection systems are capable of interfacing with a GIS, the extent of this integration can have a major impact on usability. For example,

ArcGIS will enable users to attach other files to a shape file – work orders, PACP inspection reports, video data, etc. – such that when someone calls up the shape file, they can see this associated information. This is certainly useful – being able to click on a problematic line and see the most recent survey can be a tremendous timesaver. However, true GIS integration merges the two data sets even further.

PACP inspection data indexes defects within a pipeline in terms of both the distance from the start manhole and the location in the circumference of the pipe. If this information can be translated into points along the shape file, locations of cracks, root intrusions, and other features could be shown with spatial precision. When analysts and engineers are considering repairing a pipeline, seeing exactly where defects lie can provide valuable context for a repair. For example, traffic control can be planned much more accurately if a defect can be located precisely within an intersection.

Illegal connections can also be quickly tied to property ownership data. If dye test results are plotted along the length of a sewer shape file, and taps from 9 to 12 o'clock are placed on the left side of the street and taps from 12 to 3 o'clock are placed on the right side of the street, offending homeowners can be quickly located and notified by correlating this information with property records. There is no need for a crew to confront a homeowner when this data can be easily cross-referenced behind the scenes for formal notification.

GIS also makes it possible to evaluate consequence of failure. While many traditional consequence of failure calculations focus on levels of service interruption and key customers, failure location can also play a significant role. If a municipality is planning to pursue leaks in multiple water mains that acoustic sensors have identified as problematic, it may choose to start with mains that are located under busy streets, or in close proximity to hospitals, as catastrophic failure of these mains could cause property and economic damage that goes well beyond service disruptions.

One question that often comes up when updating GIS information is how to treat pipelines with bends or curves that may not be well documented. Simply creating a vector between the start and end manholes would be woefully insufficient in such a case. 3D lidar inspections can help solve this problem by producing a bend radius report, but remember, these reports can have significant sources of error due to uncertainty in lidar measurements. A more robust tool involves the use of specialized gyroscopic mapping tools. These wheeled cylinders typically contain a series of multiple accelerometers, gyroscopes, and IMUs, and are pulled through a pipeline, providing accurate meander and incline measurements as they travel. More consistent speeds lead to more accurate results, and this data can be used to directly create a shape file. While not 100% perfect, gyroscopic data collection tends to be much cheaper than survey-grade total station approaches, and can be paired with GIS data to reduce any residual errors. Reduct's Pipemapper is one of the leading products of this type (which has also been rebranded as the Cues AMP).

From a municipal perspective, the best way to approach GIS is with a sense of collaboration. Each department will need to tie into and utilize the same GIS system in order for it to be effective, so it is worth asking about best practices and workflows before making major software decisions. Both Granite and ITpipes support native, bidirectional integration with ArcGIS systems, but each does so in a different way.

Ask your GIS staff to evaluate any approaches, and don't discount workflow – cumbersome processes tend to get overlooked, leading to outdated maps and interdepartmental frustration. Remember, if you don't have a GIS system, some tools like RedZone's ICOM offer a built-in GIS that could serve as a starter package for some municipalities.

Finally, it is worth noting that while ArcGIS is the dominant tool used in government today, other alternatives exist. QGIS is a free and open source alternative to ArcGIS. For smaller municipalities interested in exploring the value of GIS data, QGIS can give a clever technician the tools they need to load multiple layers of shape files and begin drawing conclusions. Beware, however: while QGIS is open source, it does not have the same robust development and partner ecosystem that surrounds ArcGIS. It will be much easier to find an engineer or consultant with ArcGIS experience and to obtain a software package that natively interfaces with it. Consider this before investing time into QGIS-exclusive features.

ENTERPRISE ASSET MANAGEMENT

While condition assessment surveys may lead to localized rehabilitation programs, these cost estimates can be difficult to generalize: if one section of trunk sewer is in need of rehabilitation, will the rest of the line require overhaul? If overhaul is recommended, will costs be similar? While smaller agencies may be well served by the combination of CCTV inspection management software, Excel budgeting tools, and GIS, larger agencies that have a multitude of assets often require a more comprehensive solution. Enterprise asset management software is able to calculate the total cost of ownership of physical assets (including buried infrastructure), track work orders, calculate depreciation, manage materials and procurement, and integrate with accounting and finance systems. Many popular EAM systems have also incorporated the functionality of computerized maintenance management systems (CMMS) and can schedule crews, work orders, and optimize routes for efficiency. While there is technically a difference between EAM software and CMMS tools, many use the terms interchangeably, even if – strictly speaking – CMMS tools are more limited than EAM packages.

Condition assessment programs are intrinsically linked to cost estimates, as the timing of maintenance programs has a major impact on the overall cost of maintaining infrastructure. Anecdotal evidence supports the notion that deferred maintenance increases the overall cost of operating a complex system. For example, ignoring a vibrating or squealing noise in an automobile can often lead to a hefty repair bill if the car breaks down on the side of the road – as can happen if an aging timing belt snaps on an interference engine. Similarly, as pipelines age, trenchless rehabilitation technologies remain an option up until the point of failure – at which time costly excavation and emergency repairs become unavoidable. Interestingly, only a few have attempted to quantify this relationship.

A Dutch Engineer, Reinhold de Sitter, observed the fact that costs for maintaining concrete structures increased exponentially once corrosion began to set in and advance. His "Law of Fives" stated that structures could be reasonably monitored for a very low cost – say $1 for ease of computation. Maintaining concrete structures to prevent corrosion could be performed for a similarly low cost – $5, which was more than monitoring, but less than aggressive repairs. If that maintenance was deferred and the concrete structure began to show localized signs of corrosion, it would take $\$5^2$ or $25 to repair damage in a specific area. If those repairs are also deferred, it would cost $\$5^{2+1}$ or $\$5^3 = \$125$ to begin repairing more generalized damage to the structure due to corrosion. Perhaps more significantly, if the price of the base maintenance is $10, then de Sitter's law shows costs of $100 and $1000, respectively, for repairing localized and generalized damage due to deferred maintenance.

Axioms such as de Sitter's law are useful rules of thumb that accurately reflect the manner in which the costs of maintenance programs can rapidly escalate when put off for even short periods of time, but these formulae do not take into account the detailed costs associated with specific rehabilitation programs. In the 1960s and 1970s, mainframes and minicomputers could be used to keep track of job costing, work orders, and repair programs, but limited access to these systems prevented analysis from reaching beyond departments and limited the individuals who were able to leverage data to draw conclusions. It was this idea combined with the rise of connected desktop computers that fueled the rise of EAM platforms in the late 1980s and 1990s.

Platforms such as Hansen, Cityworks, and Maximo were created to go beyond the scope of traditional CMMS software and tell the entire story of individual assets, while simultaneously aggregating data at the municipal level (Figure 10.6). These tools provide one platform to show information about light poles, park benches, and sewer lines, and make it possible for municipalities to make data-driven decisions that could sometimes run counter to conventional wisdom. For example, a thorough analysis of data may reveal that the same level of service could be maintained by applying a root foaming treatment every other year, rather than every year. Some maintenance tasks are so illogical that the EPA estimates that between 30 and 70% of maintenance activity is misdirected and does not work to prevent failure in a cost-effective way.

Reliance on EAM software increased with the recession of 2007, as reduced revenues forced many state and local governments to make a series of difficult decisions. Freezing hires and budgets left many agencies struggling to find ways to maintain an appropriate level of service as infrastructure continued to age, and inflation gradually increased costs of materials and services. While deferred maintenance was already a problem going into the recession, it became even more significant as spending slowed, modernization projects were delayed, and agencies struggled to adjust to demographic changes. Wasting time and effort on ineffective maintenance operations was no longer excusable.

The Municipal Research and Service Center estimates that a public works department could save up to 20% in overall maintenance costs by using an EAM appropriately. As a result, these software packages have become a part of government operations in every state – linking condition assessment and work order history to capital planning and budgeting. More than a dozen software packages performing this functionality currently exist on the market today. Some are supported by major tech giants like Oracle and IBM, while others are tightly integrated into other

FIGURE 10.6 EAM systems are designed to manage all aspects of a municipality's assets – ranging from physical to human resources. Many systems provide a variety of continuing education and training programs to help customers get the most out of software and to enable developers to add new features. This photograph from the CityWorks(X)po is from 2014 and was provided by VWCC Media Geeks. Creative Commons: https://ccsearch.creativecommons.org/photos/b3d8d785-13fb-4719-8e11-cd9fedadde3d

product lines, such as Cityworks, which is now a part of Trimble. Other systems are produced and sold by much smaller companies such as iWorQ.

While a variety of EAM systems exist, all seek to accomplish some of the same goals. By integrating data into a single system, information from formerly disparate activities can be leveraged for unified decision-making. For example, while age is a useful proxy for estimating service life, not all assets deteriorate at the same rate. If the costs of each work order for repair activity are automatically attached to a specific pipeline, a much more accurate list of prioritized repairs can be generated. While it is possible to link this data manually, a well-configured EAM ensures that it happens automatically, every time. Further, activities from multiple departments can be coordinated. If a major road is going to be resurfaced, it may make sense to perform lead service line replacements a month earlier so as to not tear up a newly paved surface.

While well intentioned, individual departments often do not think in these terms, or rely on informal relationships in order to communicate with others. EAM platforms take this positive communication and formalize it into something that occurs whether or not individuals think to do it. From a condition assessment perspective, it is critical that EAM systems are able to easily ingest data from inspection operations – and do so in a way that allows data to be leveraged. It is much more useful to link a PACP quick score to an EAM so that replacement algorithms can take it into account than it is to link a PDF report or video that is difficult to analyze.

EAM systems generally come in two main configurations. Hosted systems run on a server inside the IT department of a municipal government. While the vendor may install and set up the program initially, it is maintained by dedicated IT staff – for better or worse. Because it is hosted locally, updates can be tested and deployed on a regular schedule, and IT staff can quickly respond to any issues that arise. The downside of premises hosting is that local issues such as power failure or disaster (fire or flood) can take the entire system offline. Municipalities often have the option of paying a lump sum to install and configure the platform, and a smaller monthly fee for access to updates and patches.

Cloud-based or software-as-a-service (SaaS) platforms are typically hosted by the vendor in a remote location and are accessed via an Internet connection. Because the system is hosted by the vendor, all updates and patches are taken care of, but some issues may take some time to track down, depending on the level of support. For example, if the vendor upgrades a module that causes a problem for a municipality, rollback might occur over a 24-hour period, rather than an hour or two if the local IT staff discovered the glitch. Further, broad service outages at providers like Amazon Web Services can result in unexpected downtime. Sometimes these systems can be provisioned with a low upfront cost and a monthly fee to access the software, usually applied per user account.

While hosted systems dominated municipal IT departments in the 1990s and 2000s, they have been almost completely replaced by cloud software. Vendors make upgrading to cloud systems more and more attractive (as monthly subscription fees translate to predictable sources of recurring revenue), and support costs for centralized cloud infrastructure are much lower than supporting a patchwork of users, each running a slightly different version and configuration of a software platform.

Unfortunately, cost and learning curve tends to preclude smaller municipalities from taking advantage of EAM systems. While some SaaS EAM platforms can be licensed at less than $170 per user per month, in order for that user to be productive, they must be trained on the software – an expense that can range from a few hundred to a few thousand dollars. Data can be migrated into the system as well, either by manually ingesting thousands of spreadsheets, or by paying a professional services fee to import legacy data. These import costs can range from a few thousand to potentially tens of thousands of dollars depending on the complexity. Any customization such as linking with other systems or software packages can also require a fee of several thousand dollars to establish. For small to medium municipalities, budgeting at least $10,000 to set up and configure an EAM platform is reasonable.

After going through the process of configuring a new system, it can be daunting for municipalities to contemplate switching to a new vendor. Not only would data have to be migrated from one system to another, but with cloud-based systems, there can even be a fee for exporting data from the existing system – remember, the servers are controlled by the vendor. Consider including a clause in contracts that require a cumulative export of data to be made available at no cost. While larger vendors may be unwilling to accept these terms, this can be used as a negotiating point to reduce service fees.

Choosing an EAM system can be a daunting exercise – with so many systems available, how can a municipality ensure that it is selecting the right software platform? One source of information is nonprofit industry groups or trade associations. In 2016, the Water Finance Research Foundation, a nonprofit group dedicated to researching best practices in water infrastructure, evaluated 14 major EAM systems across multiple criteria including overall cost, support agreements, condition assessment tools, risk management utilities, GIS integration, and support for future expansion. From a purely functional perspective, Cityworks, Oracle, Cartegraph, Maximo, and Infor (formerly known as Hansen) were the top five platforms. However, when cost was taken into account, Cityworks, Cartegraph, Energov, Lucity, and Oracle were the top five in terms of value for a given set of features.

Be careful – when searching for EAM platforms, a wide variety of inapplicable products may turn up in search results. The term 'EAM' can also be used to manage the lifecycle of a variety of assets in multiple industries – from building construction to IT infrastructure, to mining and heavy equipment. While the premise of EAM is the same across all platforms, it is worth using software that is designed for municipal use, as features are already targeted toward that user base – meaning much less customization is required to make a tool productive.

While most water and wastewater users will not be placed in the position of choosing a citywide EAM system, many will have the opportunity to make more localized decisions, such as which CCTV inspection platform to use, or what type of acoustic monitoring devices should be installed. When selecting these tools, it is critical to weigh integration into the CMMS as a factor to consider. If a tool is easy to use day to day, but requires extensive manual involvement to transfer data into the EAM, it is ultimately not the right system for the department. Fighting the EAM requires time and energy, and is ultimately a battle that will be lost. Make EAM integration a condition of platform adoption, and don't settle for bare-bones functionality. Insist on useful integration of data – if APIs are accessible, require that vendors use them, rather than a manual import–export process.

If the EAM is a source of frustration, get creative. Use your preferred CCTV software package to plan, schedule, and route crews, but make sure that the package can communicate bidirectionally with the EAM. Create the work order on the EAM, read it into the CCTV software, complete the required inspections, and upload the results back to the EAM. Don't create an orphan process that parallelizes EAM functionality – make sure data starts and ends in the system of record.

Finally, many software vendors eager to get municipal business will promise EAM integration as a condition of contract award – there is not necessarily anything wrong with this, as every vendor needs to start somewhere. Don't pay a privilege for beta testing and insist on a service-level agreement (SLA) if you are paying full price for the service – require that bugs be fixed quickly and to the standard of the municipality. If the EAM vendor requires certification for interoperability, put that onus on the vendor. Don't risk a warranty or service agreement by enabling unauthorized programs to tap into municipal systems – even if well intentioned.

11 Future Trends in Condition Assessment

Previous chapters have given an overview of the current state of the art in many types of condition assessment tools for buried assets. Technology, however, continues to improve, and the standards and practices listed will continue to evolve in the coming years. City planners and municipal officials can anticipate some of these changes, as many of them are on the precipice of commercialization and widespread attention. Others are gaining momentum slowly and steadily, improving incrementally year over year. While it is impossible to say with certainty what the condition assessment landscape will look like in 10 years' time, this section contains some topics to watch that are likely to have an outsized, disruptive impact on the space.

In one sense, water and wastewater tends to be a small community, so it is usually worthwhile to check in with key individuals to see what they are working on from time to time. In another sense, many innovations will seem obvious to those in the broader technology field, as they represent the trickle-down effect of popular technologies. Advanced image processing and parallel computing can seem firmly planted in the domain of futuristic autonomy and supercomputers, but over time, the cost of these technologies decreases and they become more and more available for broader applications.

While the specific technologies listed may not be relevant to every facet of water and wastewater infrastructure, many are broadly applicable, and it is worthwhile to think about current pain points and what a technical solution to those pain points could look like. Perhaps one of these technologies could be slightly modified to solve a pressing internal problem, or perhaps completing the thought exercise will help refine what to look for at trade shows and in industry publications. Remember, early adopters often get to try out a new technology at no cost in exchange for feedback or press, so don't be afraid to reach out to a company or individual doing something interesting.

5G AND CONNECTED PLATFORMS

Next-generation wireless technology – or 5G – is already transforming multiple industries. While the term "5G" simply refers to the fifth generation of digital cellular networking technology, the underlying technology represents a significant improvement over existing wireless standards (Figure 11.1). Industry groups have termed 5G "New Radio" (NR) as it operates on two different portions of the electromagnetic spectrum (including bands above 24 GHz – a first for cellular technology). Strategic use of these bands enables 5G to rival wired fiber-optic connections in many ways.

FIGURE 11.1 The deployment of 5G and other advanced network technologies will enable devices to be easily connected to a broadband network without requiring expensive wiring. This photograph shows a Vodafone tower with an added 5G antenna (boxed in red). Image provided by Tomás Freres. Creative Commons: https://commons.wikimedia.org/wiki/File:Vodafone_5G_Karlsruhe.jpg

First is the provision of high bandwidth communication – meaning large files can be rapidly streamed or transferred. 4G technology can theoretically provide download speeds of 450 megabits per second (Mbps), but 5G NR will increase this by more than a factor of 10. Download speeds of 10–20 gigabits per section (Gbps) will be possible, and this rapid data transfer ability is termed enhanced mobile broadband (EMBB). Not only are those speeds faster than current Wi-Fi technology – they are faster than the Ethernet links installed in many buildings, and are even faster than the SATA interfaces used to internally interface hard drives with computers (Figure 11.2). EMBB has the potential to replace wired Internet connections: while a fiber-optic line into city hall was previously the gold standard of bandwidth, a 5G connection could be faster, cheaper, and easier to upgrade without digging up physical infrastructure and installing miles of new cable.

EMBB means that every device can have a high-speed Internet connection at its disposal, just as though it were plugged into a wired network or cable modem. Accordingly, field technologies incorporating 5G will be able to seamlessly transfer large files without physically connecting to a network backbone. CCTV trucks will be able to enable a wireless modem and transmit a day's worth of footage back to a server in the office much more quickly than plugging in a USB drive. Sensors can be configured to transmit in high resolution at all times – security cameras can send full 4K 60fps videos continuously, making it much easier to identify someone

Mobile Generation	Usage ID	The range of Frequencies; (Examples)	User Data Bandwidth (Practical examples)	Coverage per Antenna & usage
3G	Mobile	850MHz, 2100MHz	2-10 Mbps	50 – 150km Suburban, City, Rural area
4G	Mobile	750MHz, 850MHz, 2.1 GHz, 2.3GHz and 2.6GHz (Centimeter wave)	10-30 Mbps Long-Term Evolution (LTE) version	50 – 150km Suburban, City, Rural area
	Fixed Wireless		50-60 Mbps Long-Term Evolution (LTE) version	1 – 2km Home, office and high density area
5G	Mobile	3.6 GHz, 6 GHz	80-100 Mbps	50 – 80km Suburban, City, Rural area
	Fixed Wireless	24-86 GHz (Millimetre wave)	1-3Gbps	250 – 300 m Home, office and high density area

The summary of Frequency and data bandwidth: 3G to 5G cellular mobile generation
- 5G network real-world test: examples by Qualcomm's simulated 5G tests on Feb/2018

FIGURE 11.2 Table showing the available user bandwidth for various cellular technologies – ranging from 3G to 5G. Notice that the available bandwidth for some frequencies of 5G will enable wireless connectivity faster than wired gigabit Ethernet. Image provided by Goodtiming8871. Creative Commons: https://commons.wikimedia.org/wiki/File:Frequencies_and_data_bandwidth_-_3G_to_5G.jpg

attempting to access a manhole without permission, than by looking at the choppy, low-resolution video that is state-of-the-art today.

This can be extrapolated to other types of sensors, as wired data connections are no longer necessary. Many current devices tend to bundle the transmission of data into packets or chunks for transmission back to a server for analysis – think acoustic leak detection sensors that report their status every 12 hours or so. When provisioned with appropriate batteries, SmartCover level sensors, acoustic hydrophones, and flow meters could continuously record and transmit data, potentially reducing incident response times by hours or days. New sensors could be installed to take advantage of continuous transmission – such as mini zoom cameras installed in problem manholes around a collection system – spotting problems in real time, as they develop. In the past, it would have been almost impossible to transmit a high-resolution video from a network of cameras over a cellular connection in real-time, but 5G has the promise to make it the new norm.

While 5G NR will enable the water industry to engage with much larger quantities of data much more quickly, it will create new bottlenecks on the back end. Just because devices can generate large quantities of data does not mean that servers are equipped to store that data, and that computing resources have been provisioned to process the data. If a camera captures footage of a surcharge occurring, what good will that do if data is discarded so routinely that nobody is able to see it? Municipalities should think about cloud or hybrid storage systems – especially systems with built-in deduplication – to support this next generation of high-data sensors.

The 5G standard also contains a specification for massive machine type communications (MMTC), which is intended for the growing industry of things. A major reason why sensors bundle data into packages for transmission is that wireless radios tend to use a significant amount of power, and conserving battery life can require strategically operating antennas and modulating transmission accordingly. MMTC is explicitly designed to solve this problem, albeit at the expense of some bandwidth. For sensors that do not require Gbps, this can provide an ideal way to increase data transmission while conserving battery life – for example, flow meters generate a few kilobytes of data every second. Meters adopting MMTC could transmit this data in near-real time and still dramatically increase battery life. MMTC is not the only technology that promises to maximize battery life for low-power communications. Some technologies such as LoRa can transmit over extended distances, while others enable peer-to-peer communications and mesh networking. Keep an eye on Internet of Things (IoT) and Industry 4.0 technologies for advancement in this space (Figure 11.3).

While it is clear that 5G will be easily applied to sensors throughout a water and wastewater system, surprisingly, the technology can also be used for actuators. An additional feature of the specification calls for ultrareliable, low-latency communications (URLLC), which focus on the response time rather than the overall speed. To envision this relationship, imagine a streaming video conference call. High bandwidth communication means that the video looks crisp and fluid and pixelation is a rare occurrence. Low-latency communication means that there is a minimal delay between the speaker saying something and the recipient hearing it – there is no awkward lag in communication, where the parties tend to talk over one another. While low latency has been present in wireless communication for some time, it has yet to be guaranteed in this way. With URLLC, instructions can be received between 1 and 4 ms after transmission, meaning electronically controlled valves and weir gates could be linked to a control system via a wireless 5G connection.

While it can be unsettling to imagine connecting these devices over a wireless connection, the lower cost of doing so means that more and more devices can be linked into the system. Security concerns become paramount when actuators are linked into an overall network – remember, some viruses and malware have explicitly attacked SCADA systems. However, a well-integrated network of sensors and actuators could produce a truly "smart" water system. Imagine a system that detects a small leak in the middle of the night, reduces flow to the affected area, and automatically schedules a crew to the affected area at the start of the next workday. These kinds of feedback systems can be enabled by linking sensors and actuators together with 5G.

While 5G is a vast topic, one additional item that may be of interest to municipalities is the provision for a citizens broadband radio service (CBRS). CBRS has no relation to the older citizens band radio, still used by some long-haul truckers, but instead represents a portion of wireless spectrum that is lightly licensed, and uses a dynamic spectrum access system to manage frequency allocation among different entities simultaneously using the spectrum for transmission. Effectively CBRS makes it possible for interested parties to deploy their own 5G or LTE systems – building out well-controlled networks that cover large areas that would have been

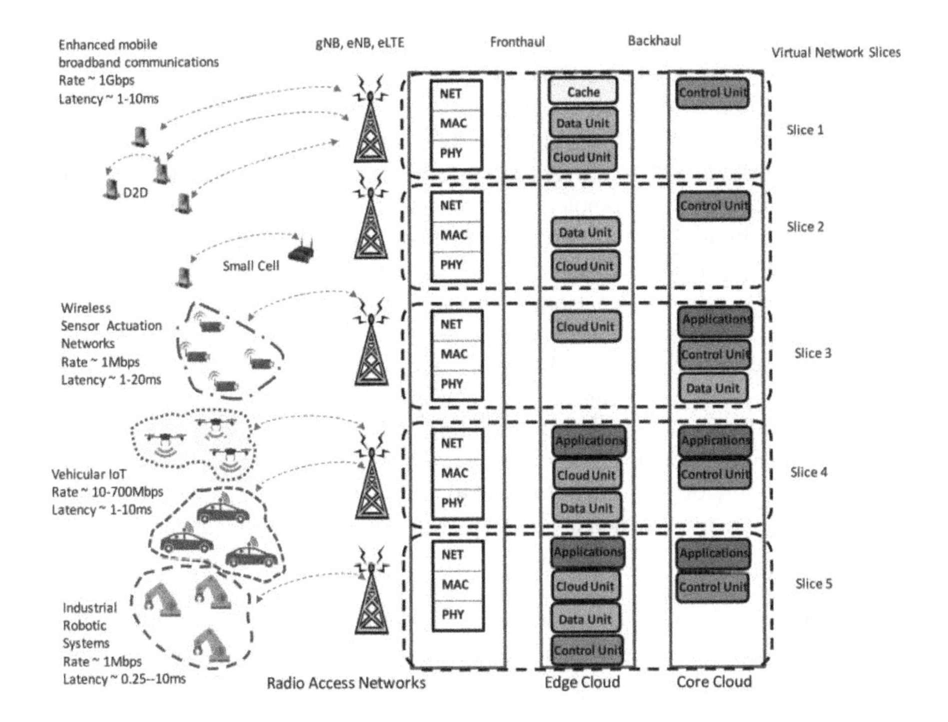

FIGURE 11.3 The term "5G" encompasses many different technologies – ranging from high-speed connections for traditional cell phones, to low-latency connections for actuators, to low-power connections for the embedded Internet of Things. This graphic shows some of the possible architectural configurations and was taken from Akkari and Dimitriou, *Computer Networks.* STM (Elsevier): https://www.sciencedirect.com/science/article/pii/S1389128619306346#fig0002

ill suited to Wi-Fi. Cellular infrastructure has always required some level of dependence on carriers. CBRS would reduce that and give municipalities the option to devise a way of "rolling their own" network – something that could be appealing if initial 5G pricing is too high.

Making the leap from water and wastewater to cellular infrastructure can be challenging. Companies like Twilio offer platforms that make it easy to explore the notion of connected devices, without diving into baseband software and firmware programming – this can be helpful if you have existing devices that you would like to interconnect, but don't have the level of vendor support required to do so easily. A consultation with a wireless expert about future plans and goals could prove to be extremely worthwhile – Oku Solutions and other firms can help advise municipal governments on standards, build-outs, and the right way to phase in technological adoption.

It is also important to acknowledge that existing tools continue to integrate with mobile devices using current wireless technologies. Many EAMs can link to mobile device applications – verifying that crews are at the right location, timing the length of the service call, and accurately reporting on maintenance information that would have been manually logged in the past. While some regard this as invasive (the EAM is "checking up" on employees – and this may present a labor relations issue), written documentation tends to be a significant source of error, especially when completed in the field. Manually logging the start time, stop time, and all equipment used in the middle of a rainstorm becomes an exercise in approximations, especially if attempted after work is complete. Automatic links to mobile phones can lead to much more accurate data collection. Rather than estimate a start time of 9:30, the mobile device can show that the crew arrived on site at 9:28.

Use caution: excessive dependence on mobile devices can lead to interruptions in productivity if a device is lost or damaged – something that is inevitable if the devices are used in fieldwork. Field phones will be dropped in manholes, run over by trucks, and suffer more damage than consumer devices. Plan for this eventuality – back up all user data to the cloud, and provision a standard phone. Remaining a generation or two behind can allow each team to literally have a box full of extra phones in a supply drawer, ready to be activated and reprovisioned in the event of disaster. Remember, municipal applications will not require the latest processor, so standardizing on an older device will likely be acceptable. Check with software and app vendors accordingly.

While nobody likes carrying around two phones, use caution when enabling field staff to use their own devices. People tend to use a variety of mobile devices – some use Android, some prefer iOS, and many continue to use older devices well beyond the point of vendor support. Any savings projected from a BYOD policy can quickly become offset by increased support costs that stem from trying to make an application work on a version of Android that has not been supported for several years. Further, while the risk is remote, employee devices could serve as attack vectors for municipal systems, especially if unpatched and connected to critical infrastructure. Many unscrupulous applications have been found to siphon data from other areas of the phone, log keystrokes, and perform other malicious activities. If one of your employees exposes their credentials to critical systems, miscreants could use this information to stage an attack. One frustrated resident could wreak havoc with just a few minutes of remote access to a SCADA control system.

BIG DATA AND ARTIFICIAL INTELLIGENCE

The vast quantities of data that will be generated by a network of connected sensors can be processed and analyzed using some of the Big Data technologies discussed previously in this text. However, large quantities of data are also forming the core of artificial intelligence applications, which have the potential to revolutionize multiple aspects of condition assessment – removing human subjectivity from analysis and codifying information in a standardized way.

While AI is commonly associated with robots and science fiction, some of the core technologies are relatively simple to understand and have already been implemented in everyday objects. For example, AI based on computer vision and image processing is used by iPhones as part of the "Face ID" function, where the device will be unlocked when an authorized user's face is visible in the camera. This isn't

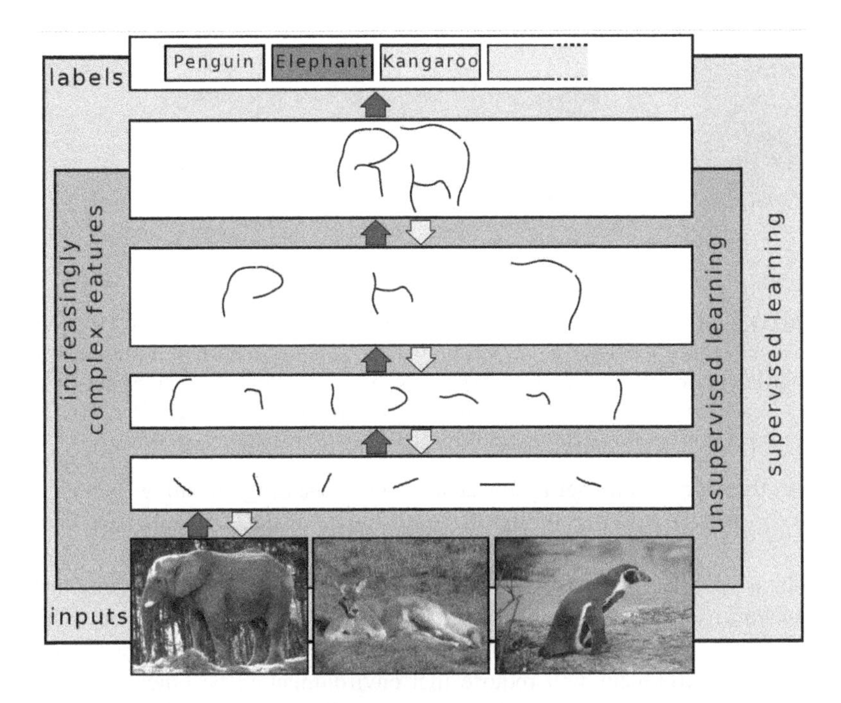

FIGURE 11.4 Deep learning is one part of artificial intelligence than can be used to train systems on complex tasks. The value in deep learning systems lies in both the algorithms and the training data, so hoarders of old inspection information may be rewarded. Diagram by Sven Behnke. Creative Commons: https://commons.wikimedia.org/wiki/File:Deep_Learning.jpg

just pattern matching – the system understands what makes a user's face significant – even when obscured by facial hair or glasses, or when presented in different lighting situations. AI enables the system to develop a more detailed picture of what the user looks like by processing more and more data over time (Figure 11.4).

Many systems that apply AI to image processing rely on underlying systems called convolutional neural networks (CNNs). Neural networks are computer algorithms that are based on the structure and function of the human brain – in which interconnected neurons enable complex processing and reasoning. Each neuron is simple – high school biology students learn about the action potential, and the triggering thresholds – but connecting millions or billions of neurons together (the human brain has approximately 100 billion) can facilitate complex problem-solving. Teaching these neurons the significance of different patterns of firing is the essence of learning (Figure 11.5).

Surprisingly, graphics processors have enabled the effective deployment of neural networks in modern devices. Central processing units (CPUs) in a typical computer tend to be complex devices, capable of processing a variety of instructions. A CPU can add, subtract, multiply, or divide, and most chips are comprised of smaller numbers of functional units that can perform a range of mathematical computations.

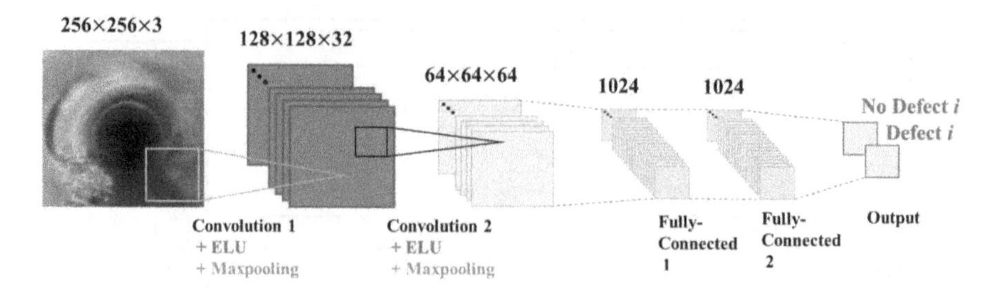

FIGURE 11.5 Deep convolutional neural networks can be used to analyze CCTV inspection data and automatically recognize defects in images. These systems have been described in academic literature and are beginning to be commercialized through ventures like SewerAI. Image taken from Kumar et al., *Automation in Construction.* STM (Elsevier): https://www.sciencedirect.com/science/article/pii/S0926580517309767#f0020

Further, they are usually set up for continuous processing, meaning they are optimized for serially working through large quantities of data in a relatively complex way.

Graphics processing units (GPUs) are structured very differently. While humans are trained to wait for the results of a CPU, the GPU is typically tasked with generating visual information in near-real time. In a video game, a character moving through an environment will require that environment to be quickly updated in response to actions and presented with high fidelity in order to be believable. To accomplish these goals, many GPUs consist of thousands of very simple compute units, each of which is dedicated to a specific task. Different vendors call these units different things – NVIDIA refers to them as CUDA cores, and AMD has termed them stream processors. Vendor-agnostic texts often refer to them as shaders (though this term is not ideal, as it introduces ambiguity between software algorithms and physical hardware).

Whereas modern CPUs may have 4 or 8 cores designed to process information simultaneously, GPUs have thousands of shaders, which are capable of processing large quantities of data in parallel. High-speed interconnects enable the results of these calculations to be quickly shifted to other locations in the graphics pipeline. While initially this structure was introduced to enable the rapid rendering of scenes, it has subsequently become useful for more generalized data processing. By breaking tasks down into simple instructions and executing those instructions in parallel, GPUs can far exceed the performance of CPUs for some tasks – such as modeling protein folding, or mining bitcoins.

A chip containing a large quantity of interconnected shaders maps easily onto the structure of neurons in the human brain, making GPUs ideal for machine learning and the development of neural networks. In a learned response in the brain, a collective pattern of synaptic firing is associated with a pattern or behavior – such as recognizing a scent or a face. Long periods of positive and negative reinforcement are required to transfer this pattern into memory for future access. In computing, a network of nodes based on shaders or compute units works in the same general way as the neurons in the human brain. Data flows through the network in one direction, such that nodes receive input from previous nodes and pass output onto subsequent nodes. As the system learns details of significance, nodes will assign weights to its various inputs, multiply those numbers together, and only transmit an output if the result exceeds a required significance. The goal is to feed an image into the network and have it output a value that agrees with the correct classification of the input.

Actually learning the appropriate combination of inputs and outputs that results in the correct transformation is exceedingly difficult, and computers will often try random combinations of weights in order to ultimately arrive at an optimal solution. For this reason, neural networks are often trained on cloud systems or supercomputers, because millions of permutations can be examined in a short period. Once the right combination of weights is established, a neural network can be deployed on a commodity GPU – several are available in an embedded form factor for less than $100 (NVIDIA's Jetson platform is popular among developers).

How can systems determine if the neural network is the correct match to classify images appropriately? The answer is data. In order to determine which random weighting between interconnects is appropriate, neural networks are fed large quantities of labeled images in order to train the system. Each time the system successfully identifies an image, the tested pattern of nodes and weights gains validity, and if a system is able to accurately identify hundreds or thousands of labeled images, it might be considered acceptable for general use. Larger amounts of training data result in better neural network performance. For example, if a neural network is trained on millions of images, it will tend to produce more accurate results than a network that is only trained on a few hundred, though overfitting can be a problem.

Actually obtaining labeled training data can be difficult – many of us have unwittingly participated in some of this labeling when we identify crosswalks or cars as part of an online CAPTCHA. The iPhone builds its training data over time – every time a face is not recognized, but the correct passcode is entered, it uses that new image to further refine its model – meaning that Face ID can get more accurate over time. If a user begins wearing glasses, they may have to type in their passcode for the first few days after they begin wearing them, but will notice that this frequency goes down in the ensuing weeks, as the neural network learns their new appearance.

Neural networks used in image processing are often referred to as deep convolutional neural networks (deep CNNs). While a mathematical description of deep CNNs is beyond the scope of this text, convolutions essentially are functions that break images down into smaller chunks and attempt to extract baseline features from those small chunks – such as edges. The term "deep" means that a single network could have dozens of these convolutions employed in the overall modeling. Images can be broken down into multiple dimensions and useful information or features can be extracted from each dimension. Training determines which of these features actually are significant – for example, is there something in the red space of an image that can help identify a crosswalk? If so, what is it and can it be found in other representative images of crosswalks?

While the above may sound daunting, the incredible thing about these deep CNNs is that one does not need to be a computer scientist to understand and leverage the technology. Packages that can abstract this learning can be freely downloaded and incorporated into various workflows. You Only Look Once version 3 (YOLOv3) is an object detection platform that can be harnessed by amateur developers. With a few hours of work, YOLOv3 can be trained with image data and deployed on a video stream, automatically recognizing objects. Teachers completing a workshop with NVIDIA are able to train a neural network to take attendance in their classes by scanning the room and recognizing student faces. This example is not intended to demean educators – instead it is intended to show that AI tools are becoming democratized to the point where talented individuals in other fields can leverage these tools to begin drawing conclusions – without a master's or doctorate in computer science.

Applying this AI to image processing could result in a system that could automatically process a video stream, detect defects, and assign them the appropriate PACP codes. Operators could simply focus on getting a camera to complete an inspection with high-quality video, and know that an algorithm could automatically take care of the coding after the fact. While some firms such as SewerAI have begun to demonstrate the viability of this technology, it is likely to reach its full potential when coupled with 360° imaging technology. Even if the software can detect a crack while a crawler is driving by, a more detailed view of the crack (obtained via PTZ) is essential in deciding whether or not rehabilitation is necessary. Engineering analysis cannot occur without adequate contextual information. That said, expect this technology to rapidly improve in the years to come; there may come a time when PACP coding becomes more of a QA check on an algorithm's output rather than a manually intensive task performed by humans.

The power of neural networks is not restricted to image processing. Conclusions can be drawn from a variety of municipal data sources. Neural networks enable big data to be effectively mined for interesting relationships and conclusions. For example, feeding a neural network thousands of work orders and overall project cost information may generate surprising conclusions. Perhaps one team tends to complete projects faster and at lower cost than any others. When that team is examined over time, performance can be linked to a key individual – whichever team he is on seems to experience a benefit. While AI cannot detect exactly what it is that he does, it could set the stage for an interview, observation, or training program to try to isolate the source of his efficiencies.

While this technology sounds fantastical, it is not – AI and Big Data are firmly grounded in reality, even if they have not yet crossed over into the municipal space. Entities that tend to do well with these new systems are entities that have large, well-organized data sets that can be fed into a neural network. If your work orders are on paper, and CCTV inspections are on VHS tapes, now is the time to start digitizing those materials, and inputting information into a standard set of databases or spreadsheets. When your municipality is ready to try AI, you will be poised to immediately begin generating useful conclusions if data is prepared and ready to go in advance.

This also leads to a thorny question that is seldom discussed in municipal circles – who exactly owns condition assessment data? While nobody would dispute a municipality's right to their own work orders and inspection videos, if these are hosted in a cloud system, what rights does the cloud system provider have? Can they use municipal data to train their own AI systems? Many vendors actively do this and permission to do so is buried in an end user license agreement or overall terms and conditions. While there is nothing inherently wrong with this, municipalities are effectively footing the bill for AI development by paying to subscribe to these cloud systems, and feeding them with well-labeled training data. Consider demanding something in return – free access to a model or predictive analytics platform. If you have a large data set, many vendors will be willing to make concessions in order to access a valuable library of training data.

Municipalities should also consider collaboration with universities. AI is currently a hot topic in computer science and engineering departments, and students could likely create systems that could leverage municipal data in order to draw conclusions. While student projects are always a gamble, and many result in deliverables that are less than useful, they can also be a method of injecting new viewpoints into a government organization. A promising student could go on to work as a summer intern and learn more about how public works operates, generating useful code in the process.

INSPECTION HARDWARE

Many advances are also taking place in the domain of inspection hardware that is used to collect condition assessment data. Much of this is driven by larger technological initiatives, such as the development of self-driving or autonomous vehicles (AVs). Autonomous vehicles require a variety of sensors in order to collect enough data about their environment to make navigation decisions in real time. The development of these programs has improved the quality and reduced the cost of a variety of key sensors (Figure 11.6).

Lidar has already begun to benefit from these advances. Prior to the advent of AV programs, lidar units were single-channel devices that could cost thousands of dollars per unit, and communicated over cumbersome serial interfaces. The adoption of lidar by multiple AV programs quickly led to the release of units that could collect data in 8, 16, 32, and even 64 channels at the same time, leading to the construction of high fidelity maps. Further, these devices declined in cost and added user-friendly features like Ethernet interfaces to enable rapid integration into systems like sewer robots.

AV programs continue to push the development of lidar. The spinning mirror at the heart of many units has become a point of failure in automobiles due to road-induced vibrations and bumps. Many companies – such as Quantergy – are working on the development of solid-state lidar chips that use an array of lasers and detectors and have no moving parts. While there are R&D challenges to be solved in producing these chips, they have the potential to further reduce the cost of lidar units through much more efficient manufacturing, and increase the lifespan of sensors. This means that sewer inspection robots could be equipped with an array of redundant lidar chips to effectively map a pipeline in three dimensions – a major improvement from current technologies which rely on a single sensor that can easily be occluded by debris or water droplets.

FIGURE 11.6 Autonomous vehicle programs continue to push development of sensors and other technology that can be used in condition assessment programs. This image shows a Ford Autonomous Test Vehicle outfitted with multiple lidar units from Velodyne. Self-driving car programs have sped up the development and reduced the cost of lidar units. Image provided by Steve Jurvetson. Creative Commons: https://commons.wikimedia.org/wiki/File:Ford_Autonomous_Test_Vehicle.jpg

Lidar also has limitations which are major barriers for true autonomous driving. Many lidar sensors struggle with fog, snow, and other forms of precipitation, severely restricting autonomous driving in many parts of the world. While it is unclear which technology will ultimately supplant lidar – discussion ranges from synthetic aperture radar systems to terahertz imaging – whichever approach is found to be suitable will have a major disruptive impact on underground inspection. A technology that can help cars navigate in rain or mist could potentially see through fog in a pipeline, making it possible to inspect pipes in cold weather conditions without blowers. Accurately penetrating small layers of precipitation could make it possible to see below fat and grease deposits, and even return useful electromagnetically derived sensor information from below the flow line. While it is unclear if any or all of these benefits will come to fruition, underground condition assessment can expect to reap a windfall from AV-generated technologies.

Multispectral imaging systems are another technology to watch in terms of inspection hardware. Traditional CCTV cameras used in pipeline inspection capture data from the visible light spectrum. Cameras used in other applications rely on other types of light to produce different types of images. Thermal cameras pick up on infrared radiation or heat signatures, and millimeter wave cameras can penetrate clothing to detect hidden objects (and do so daily at TSA checkpoints). Multispectral cameras apply more than one imaging mode in a single device and are becoming increasingly popular.

For example, a multispectral camera could collect infrared and visible light images, and overlay one on top of the other so that a heat signature could correspond to visible features. Many have proposed that areas of infiltration or exfiltration could be detected from a change in temperature. This system would put cracks and fractures in context to see if colder groundwater is entering the system at these locations. It can also detect conditions that are hallmarks of imminent failure. If a section of concrete looks colder and appears to be corroded, perhaps that section of the pipe is thin enough that effects from the surrounding soil are visible. After a lining project, differences in localized thermal properties could provide an indicator of where to cut taps out in order to restore service. As multispectral cameras decrease in cost, they will become easier to integrate in inspection devices, as many can interface using the same hardware as traditional digital cameras.

It is also possible to view the CCTV crawler as a generic platform, onto which a variety of payloads could be mounted – including sensors that can analyze the conditions of wastewater. By mounting a pH meter or dissolved oxygen sensor onto a CCTV crawler, industrial discharges or locations of high microbial activity could be easily detected. While interesting, each of these examples is too nuanced to be of broad applicability. However, if a standard were developed for CCTV crawlers that enabled a variety of sensors to be easily attached on an ad hoc basis, it could create a platform for expansion that could be adopted throughout the industry.

For example, if a municipality wished to observe pH at each incoming tap along a line, they could ask their crawler vendor for a quote to electrically integrate a pH sensor into the system, or mount a commercial probe in a location where it could be seen by the built-in camera. A standard interface for these additional sensors could allow municipalities to buy a crawler, and mix and match sensors for a variety of applications, knowing that each would not require custom integration or ad hoc engineering. While an industry-wide standard may be unrealistic, it could be interesting for a single vendor to define an interface and invite collaboration as part of a way to differentiate their products from the competition.

VIRTUAL REALITY AND AUGMENTED REALITY

While virtual reality (VR) is often discussed in the context of video games and entertainment, it has applications in the water and wastewater space – especially now that the technology has become portable and can be easily used by individuals without being tethered to a computer. Inspections of water and wastewater lines obtained with fisheye lenses or multiple cameras can be projected into a simulated environment – enabling users to actually travel inside a pipe or manhole and look around at an area in detail.

Many view this application of VR as no more than a disgusting gimmick – a 3D simulation of a wastewater structure can show details of debris and grime in unpleasantly-high resolution. Yet, these simulations can be extremely useful for training and job planning. If an individual has never been inside a manhole before, a VR simulation can help them see features of interest and hazards in a controlled setting. This can also be useful when planning for rehabilitation. Actually viewing a defective area from multiple perspectives can help a rehabilitation plan come into focus.

Three-hundred and sixty degree cameras are not limited to crawlers or manhole inspectors – they can also be purchased as standalone devices that can be carried into a variety of settings. In this manner, virtual environments can be constructed for control rooms, pump stations, dry wells, and other structures that technicians may need to access. A well-structured VR training program can orient users to hazards in an immersive environment. Research indicates that learning in a simulated environment may be more readily recalled in the actual setting – so rather than explain a workaround procedure or checklist to a new crew, showing them the actual procedure in a virtual environment may be more effective.

It is important to note that virtual reality is not for everyone. Certain individuals experience severe disorientation, nausea, and discomfort as current VR systems wreak havoc on the vestibular system. The reason is often attributed to a slight delay between actual movement and seeing the result of that movement in the head-mounted display. Even a few milliseconds (on par with the refresh rate of the unit) can be enough to make certain individuals completely incompatible with virtual reality systems. This isn't a condition that can be overcome through learning or exposure – people who become violently ill with these systems will continue to do so until the technology improves. Therefore, consider using VR as a supplemental training program, and be prepared to provide alternative approaches for individuals who become ill when using the technology.

Augmented reality (AR) has more promise for industrial applications. In AR systems, individuals view the world through a device – such as a screen or lens mounted in front of the eye. These screens allow the user to view the actual world, but also can project computer-generated data into that viewpoint – augmenting the perspective. While devices such as the Oculus or Vive are classified as virtual reality systems, devices like the Microsoft Hololens or Magic Leap One are augmented reality units. Mobile phones can also serve in this capacity, by showing the camera view on the screen while also rendering additional information in that same space – such as a heads up display, language translation, or even Pokemon characters (Figure 11.7).

When condition assessment data is loaded into an AR system, the asset can be examined in a way that goes far beyond the confines of a screen. A team of people, each equipped with an AR device, could rotate and zoom in on areas of interest at the center of a conference room table, physically pointing at representative locations on a 3D model. This model could also be projected in alternative locations. An AR headset could enable a technician or engineer to view a pipeline as it lays under a street or sidewalk. This can be useful when visualizing major repair programs – if a structure is severely deteriorated, seeing the pipeline in context can vividly illustrate the consequence of failure, helping to convince officials to quickly approve funding.

FIGURE 11.7 Augmented reality systems show computer-generated images overlaid on the real world. This technology can be used for everything from gaming (Pokemon Go) to industrial maintenance and repair. It can enable buried infrastructure to be overlaid on the streets in their actual location – making it easy for crews to confirm they are digging in the right location. Image provided by the HAHN Group. Creative Commons: https://commons.wikimedia.org/wiki/File:Einsatz_von_Augmented_Reality_im_Service_von_HAHN_Automation.jpg

However, showing a pipeline in context can also be useful for planning a major rehabilitation effort. Traffic control can be planned, business owners can be consulted, and factors that might not have been incorporated into planning documents like overhanging trees can be observed and dealt with accordingly. While expensive headsets like the Hololens provide a more immersive experience, mobile devices can replicate this to some extent by using visual clues and GPS coordinates to overlay the image in the correct location, as well.

The combination of AR/VR and 5G has created an interesting use case known as digital twinning. In digital twinning, an electronic copy of a complex system is created and projected into a separate location. The digital twin is continuously updated so that changes that occur in the real system are reflected in the digital model. Systems outfitted with a variety of sensors are well suited to digital twinning, though image processing and high-resolution cameras can detect many features without the need for sensors.

As an individual works on a real-world system, a team of observers or supervisors could be watching their progress on the digital twin – pointing out feedback and corrective action in real time. If a pressure reading begins to increase, the team could immediately tell the technician which valve to turn in order to relieve the system. While digital twinning is possible today, 5G technology will make it possible to transmit much larger quantities of information in real time so that there is much less of a lag between actions and the results propagating through to the digital twin. Digital twinning can be extremely useful for training purposes or for maintaining expensive equipment in confined spaces – for safety concerns, only one individual may enter the confined space, but he can carry with him the combined knowledge of a team accessing the digital twin to make the necessary repairs or execute specified procedures.

Interestingly, standards for AR and VR continue to evolve. In its current form, AR is able to accurately project three-dimensional models in space – so laser or lidar scans of pipelines will be most suitable for this technology, though vectorized GIS representations can also be used, albeit with less fidelity. VR benefits from high resolution image and video data, so more megapixels will generally produce better results. Modern cameras can capture data in as much as 11K, but many devices will downsample this for display. Try to retain data in open standards such as shape files, 3D meshes, and industry standard photo and video formats that can be transformed into appropriate projections on demand. As both technologies continue to evolve quickly, capture data with as much fidelity as possible early on, and be prepared to convert files into a variety of formats to function with newer devices as they are released.

ADVANCED AUTONOMY

It is also likely that the autonomous capabilities of sewer inspection robots will continue to improve – again, leveraging trickle-down effects from AV and mining technologies. The vast majority of inspection robots in use in sewers today are either passive floating platforms, or teleoperated devices that are manually driven. RedZone's Solo robot offers a degree of autonomy – but it is limited when compared to autonomous vehicles.

RedZone's Solo robot is capable of accurately navigating a pipe segment between two manholes without intervention. The Solo uses a suite of sensors to recognize when it enters an open chamber, and other sensors to avoid moving over a drop connection or step that could cause it to become stuck. If the Solo encounters an obstacle, it attempts several strategies to dislodge that obstacle before returning back to its starting location. Almost all of the AI in the Solo is coded in the drive systems and onboard computer – it doesn't perform image processing or decide whether to turn left or right, and the overall length of an inspection is dictated by the amount of onboard tether. To prevent the Solo from getting stuck and mitigate the need for difficult (and possibly expensive) retrieval procedures, the Solo limits its autonomous travel to single pipe segments.

Even that limited autonomy can result in tremendous time savings. RedZone typically bundles its Solo robots into squads of four units. One operator can deploy a Solo in one manhole, close the lid, and move to the next manhole to deploy a unit while the previous Solo is inspecting. An effective user can manage four robots, moving from manhole to manhole deploying along the way. One person, in ideal conditions, could inspect 10,000 feet of pipeline in a single day using a squad of Solos. With this autonomy comes some drawbacks – Solo robots can malfunction, and the operator only becomes aware of this malfunction upon retrieval (and review of data), and these units are more likely to get stuck than teleoperated crawlers, as onboard autonomy cannot rival human decision-making.

While the Solo can autonomously navigate through a pipeline, humans still need to collect video data and apply PACP coding. Future robots will likely perform any analysis on the robot itself. Embedded graphics processors have become so inexpensive that they are easy to integrate into designs and can enable much more complex decision-making. An autonomous robot using these chips could code all defects in a pipeline and if it is unable to complete its mission, be able to identify the likely cause of failure – blockage due to roots, collapsed pipe, or some other factor.

Robots will also be able to navigate autonomously for longer and longer distances. Crawlers could be preloaded with a map of a system, and establish their position within the system using a combination of IMUs, high resolution cameras/visual odometry, and inexpensive lidar sensors. These devices could inspect as much of the system as their battery lives allowed, and transmit their locations to a base station on the surface using low frequency RF, or perhaps a system of antennas attached to manhole covers. One robot could be deployed in the system and retrieved hours later, after inspecting (and coding) hundreds or thousands of feet of pipelines.

Systems using long-range autonomy will likely become lost much more regularly than tethered devices. Some robots will end up at the wastewater treatment plant, while others will ride surges and become stuck on manhole benches. Occasionally losing robots will become a cost of operation, and advanced manufacturing technologies will reduce the price of these units so that these losses are tolerable. For example, a 3D printed robot with commodity imaging and graphics processing hardware could be produced for a few hundred dollars and retail for a few thousand. If the robot is lost or destroyed, a new one can be deployed from inventory – this stands in contrast to today's crawlers which can sell for tens of thousands of dollars, and require care and maintenance in order to enable longevity.

It is also possible that drones will be used to inspect sewer systems. While the idea of drones in sewers sounds far-fetched, remember that in many interceptors, ample headspace is available. Drones can be equipped with a variety of sensor packages, and many come with built-in software to avoid collisions with walls or other surfaces. Drones can inspect pipelines much more quickly than tracked or wheeled robots or even floating platforms. If an inspection only needs to collect information from the headspace, drone inspections could gather this information in a fraction of the time, reducing the costs of bypass pumping and flow control accordingly.

Some engineering will be required to make today's unmanned aerial vehicles (UAVs) suitable for pipeline inspection. First, since autonomous flight is not permitted by the FAA, commercially available drones are all controlled by a remote operator using some kind of radio frequency signal. Repeater antennas would likely be required to channel this RF signal into the manhole and longitudinally down the pipe to the drone itself. Since underground structures are not considered to be part of the national airspace, a form of autonomous flight could be possible. Autonomous navigation algorithms built in to the UAV itself could mitigate this, but may be difficult to implement in a GPS-denied environment (such as underground pipelines). Many companies are using drones encased in protective structures to inspect industrial equipment, and these technological advances may spread to water and wastewater utilities (Figure 11.8).

It may also be possible to bring autonomy to bear on infrastructure maintenance devices. While autonomous inspection platforms are useful, a device that could

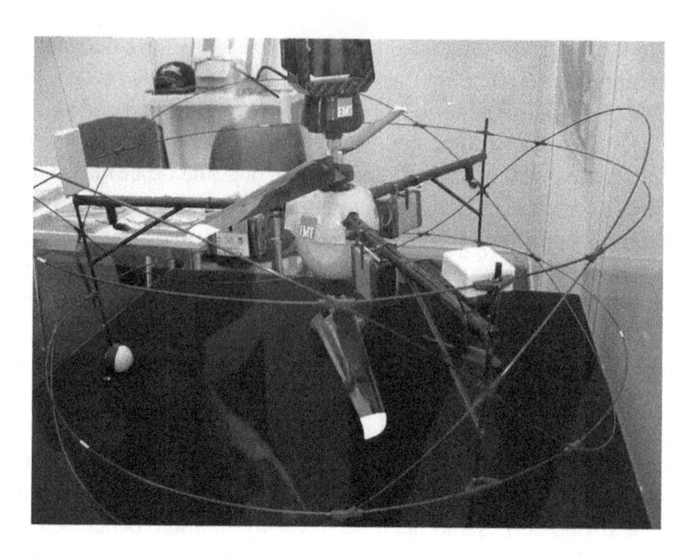

FIGURE 11.8 UAVs outfitted with crash guards could complete CCTV inspections in a fraction of the time of a conventional crawler. This image is of the EMT Fancopter and was taken at TechDemo 2008. Image provided by Swadim. Creative Commons: https://commons.wikimedia.org/wiki/File:UAV_EMT_FANCOPTER_1.jpg

analyze defects and perform field repairs automatically would be a major time-saver. If an inspection camera could detect roots and automatically deploy a cutter to remove large masses and apply a foam to inhibit growth, crew time could be spent on more labor-intensive projects. Historically, there has been a reluctance to autonomously deploy devices with cutting functionality, as it is difficult to undo a mistake – however, the right mix of sensors should enable a device with a tightly limited scope to accomplish some maintenance tasks without human intervention.

While the concept of an autonomous device roaming through the sewers continuously fixing problems may be a few years away, the idea that these devices could communicate with parts of the system is intriguing. For example, an autonomous robot could indicate its progress through the system as it passed under SmartCovers and automatically update portions of the system that were inspected in the back-end GIS, and perform multiple inspections of suspected problem areas during wet weather conditions. A robot with root-cutting functionality could communicate with the water system to trigger automatic flushing to clear debris.

SMART INFRASTRUCTURE

Nearly all of the condition assessment technologies discussed in this text involve applying some kind of external sensor to a piece of infrastructure in order to detect some kind of deterioration. This can be challenging to accomplish, as buried infrastructure is inaccessible and in many cases was not designed for sensor attachment. Clever technologies can be retrofitted into existing structures like fire hydrants or manhole covers, but other technologies must be temporarily deployed to gather information for a brief moment in time. For example, lasers must be dragged through a pipeline in order to measure corrosion on a specific date.

Smart infrastructure attempts to rectify this, by understanding that all buried assets have a finite lifespan, and technology to detect when the asset is nearing the end of that lifespan could be built into the asset itself. Imagine if the wires in PCCP could automatically notify the asset owner when they began to break – obviating the need for expensive and cumbersome acoustic installations. Such technology could be embedded in the pipeline itself and be used to monitor deterioration over the entire useful life of the asset. Deep learning and AI could even be applied to this data to heuristically model failure – giving municipalities accurate tools for estimating remaining useful life based on real-world statistics and data generated from the municipality itself.

The concept involves manufacturing pieces of infrastructure with a suite of sensors built in, which are intended to report on the health of the asset throughout its service life. A reinforced concrete pipe could be constructed with a built-in strain gauge, accelerometer, temperature sensor, electrical conductivity meter, humidity sensor, and more. These sensors could report on the continuous condition of the pipeline on a regular basis so that events leading up to failure (sudden strain) could be recognized, and the service conditions of the asset can be well understood. For example, the temperature sensor could reveal that the structure was actually exposed to several freeze-thaw cycles, despite its depth.

While such smart infrastructure sensors have been installed in bridges and dams, significant barriers exist to implementing this technology in buried assets. While 5G may solve a variety of problems for communicating with devices at the surface, it is not well suited to subsurface data links, and will quickly attenuate in soil. Thus, any of these sensors will need power and data links in order to function. Further, electronic sensors typically have a much shorter lifespan than buried assets – so if the pipe is expected to last 50 years, and the strain gauge fails after 7, why spend the extra funds to provision it in the first place? Further, would sensors from one manufacturer integrate well with sensors from another manufacturer, or could municipalities be stuck in a sort of vendor lock-in?

Thus, the original vision of smart infrastructure does not appear to be gaining traction with water and wastewater conveyance systems. However, attaching sensors to existing infrastructure has begun to gain momentum as part of a broader "Smart Cities" movement. In this movement, cities work backwards and define problems they would like to obtain data to solve, and install a system of sensors to effectively tackle the issue. For example, if a city wanted to tackle the challenge of nonrevenue water, they might install leak detection sensors throughout the system to try to reduce losses.

Many major IT vendors are providing platforms for Smart Cities that build off of EAM tools. They can sell and install a series of sensors that can attach to different types of infrastructure – from electrical outlets (to minimize power consumption) to mass transit (to improve fare collection) to sewer lines (to detect gas buildup and odors) – integrate these sensors so that data flows into the EAM, and structure capital projects that can tackle these issues by effectively leveraging the collected sensor data. Some Smart Cities programs use power savings, cost savings, and efficiency gains as a system to finance the infrastructure itself.

Use caution, however – while Smart Cities consultants can create a network of interconnected systems and ensure that planning decisions are driven by data, it is easy to see costs escalate in pursuit of perfect data. While ideally every decision would be supported by a wealth of information, sometimes the costs of acquiring that information do not justify the expected gains. Sometimes decisions will need to be made with imperfect information, and where to draw that line can require a mixture of experience and instinct.

12 The Importance of Condition Assessment

The previous chapters have outlined the many different ways in which condition assessment tools can be used to analyze and monitor the state of buried infrastructure, and it is easy to focus so much on specifications and details that the context of condition assessment programs can be lost. Remember, no municipal conveyance system is perfect. As soon as pipelines are installed and buried, they begin to degrade, and there will always be emergency calls to respond to (Figure 12.1). In that context, it is easy to adopt a firefighting mentality and move from crisis to crisis, shoring up parts of an ever-changing system.

Condition assessment programs are about taking a step back from that mentality, and creating a framework to observe and understand the system, so that gradually behavior can become more and more proactive and efficient. Remember, the EPA estimates that between 30 and 70% of maintenance activities do not effectively extend the life of buried assets. Just because a department is working hard does not mean they are making a difference and ratepayers, employees, and residents deserve to know that their system is being maintained as effectively as possible.

Without an effective system baseline, the loudest voices get the most attention. Crews will constantly respond to calls of sewage backups and odor reports, without addressing the systemic problems that may be causing the symptoms that are reported by the community. Yet, in the absence of a condition assessment program, how can one rely on anything other than reported issues? The state of the rest of the system remains unknown and it is difficult to justify investing time and resources into things that aren't a problem, especially when plenty of problems exist.

At some point, it becomes necessary to break the cycle. Water and wastewater managers need to leverage tools that already exist, and work to create a plan for building an accurate picture of the system that is their responsibility. Work backwards; if the goal is to inspect the entire system in three years, how many lines should be checked every month? Can crews spend a day or two each month proactively performing CCTV inspections, or listening for leaks at fire hydrants? Can PACP coding be integrated into emergency service calls, so that if a line is televised, its condition can be documented at the same time?

Once that initial hurdle is cleared, condition assessment programs tend to snowball, as surprising findings are uncovered. When a water or sewer line from the 1950s is found to be in great shape, while newer PVC construction is fracturing and splintering, it is natural to begin tracking down answers to those questions by gathering more and more data. At some point, a crisis will be averted, or money will be saved, and the condition assessment program will become established as a part of a more proactive departmental culture. While it can be troubling to know that

FIGURE 12.1 Photographs of two severely corroded sections of concrete pipelines. Without condition assessment programs, these pipelines could fail, causing sinkholes, property damage, and even loss of life. Photo taken from Albin, *Pipelines*. STM (ASCE): https://ascelibrary.org/doi/pdf/10.1061/40745%28146%2981

you can't inspect your entire system at once, focus on small steps that demonstrate progress.

Work backwards when designing a program. Think about the problem you are trying to solve and the data required to design an effective solution. Then select a technology that can produce the data needed to solve the problem at hand. Conferences and expositions are full of vendors who will try to convince municipalities that they need to be using a certain technology in order to prevent problems. Remember though, vendors need to sell equipment and services in order to survive. While their technology may be the right fit for your system, you know your system better than them, and any decisions should be made after research and deliberation – not after a steak dinner or an enjoyable evening at a baseball game.

It also pays to be wary of adopting technology for technology's sake – many will use advanced features as a tool to obfuscate. Buzzwords and jargon can be difficult to argue with, and nobody likes to admit that they don't understand what is being talked about. Never be afraid to ask a vendor to stop and "explain like I'm 5" how a technology works. If they can't or won't do it, how can you be certain that the solution being presented will truly solve a problem? Trust your colleagues – never be afraid to ask for references and speak to city managers or public works directors about their experiences with a particular vendor, platform, or technology.

Leverage engineers to provide guidance where appropriate – including in the drafting of contract specifications. Specifications should be designed to ensure a quality deliverable and encourage competition in a level playing field, and a sample CCTV specification is included in Appendix A of this text. Specifications provided by vendors are almost certainly designed to exclude competitors from bidding on a contract. It is acceptable to ask a vendor for input, but asking them to draft the specification will likely result in higher prices and a lack of direct competitors. Always make sure any language makes sense and can be tied to program goals. If a specification calls for lidar over a structured light laser, make sure you can articulate the reason for that. You can always disqualify a bidder if you don't like how they answer questions, but it is better to have more proposals than fewer.

Always seek examples of deliverables. If a technology is new and promising, but you are wondering how to actually leverage the data it produces, ask to see samples of said data, or even arrange for a demonstration in your system. Look carefully at both the overall quality of the data, as well as the way in which it is formatted. If videos are generated, look closely at the quality of the resulting files – are they bright enough? Sharp enough? Clear enough? Is the camera centered in the frame? Do colors appear correct? What do you intend to do with the generated files? Watch them? Link them into the EAM? Actually proceed with the process for linking a data file, and ensure it can be played back correctly. Don't accept hand waving and assurances about file formats. Comprehensive assessment programs generate large quantities of data – make sure the process works before the onslaught begins.

If you will be deploying a technology, teach crews to focus on data. For many field workers, this can be challenging, as success is clearly defined by digging up and repairing a line, or getting a robot through a section of pipe. Instead, focus on generating useful data. Reward thorough and careful inspection work, with ample attention to detail. If an acoustic monitor sounds strange, keep listening and come back the next day, rather than dig multiple holes until the problem is inefficiently located. Make sure any financial incentive is tied to data, and not another proxy metric that is easier to measure.

If you are relying on a vendor or contractor to deploy a technology, be active in reviewing submittals. Look for all elements of a specification – title cards, device speed, final report format – and provide feedback early and often. If deliverables are rejected, clearly indicate how they are to be corrected – is recollection in order or could problems be resolved in some kind of post-processing stage? Don't leave any ambiguity, and consider visiting crews in the field to observe inspection practices. Insist on seeing calibration records, and ask crews to show you the process. A crew that has never heard of the term and doesn't know where to begin should be a major red flag.

When a condition assessment system is in place, don't get complacent. Always scan for and evaluate new technologies. Subscribe to trade publications, attend conferences, and learn about upcoming trends. Seek out new technologies and try them early if possible. Watch their development over time, and be ready to replace existing programs if a promising new approach gains a financial or technical edge. Many utilities directors will lament how long it took them to move away from VHS tapes – as soon as they used digital video systems, they wondered why they didn't switch years before. For some early adopters, the answer was likely that the technology did not yet exist, but others who made the change in the late 2000s were simply behind the times.

No matter what size municipality, try to think in terms of "one government". While it is easy to get territorial and think about the needs of the water and sewer department, remember that on some level, the entire government needs to collaborate in order to effectively serve citizenry. This means working together with others to coordinate effort, minimize waste, and maintain a level of service that permeates all departments. If your city is using an EAM, get on board. Figure out ways to get your data into the system, and start drawing conclusions. It's much better to get an early seat at the table than to be the last one to arrive, kicking and screaming.

FIGURE 12.2 The aftermath of a massive sinkhole in Macomb County, Michigan. The 2016 sinkhole was caused by a failure in a sewer line, – which in turn caused damage that required more than $75 million to fix. Photo by Author.

Finally, never forget safety and the human cost of these operations. The reason why many of these technologies exist is to protect workers and provide services that make the lives of customers safer. If you lose sight of that, it can be easy to pursue technology for technology's sake, or defer necessary maintenance to meet an arbitrary budget goal. It is not overly dramatic to say that condition assessment programs can become matters of life or death. SSOs can cause illness and infection in communities, and unchecked water main breaks can create sinkholes that can cause property damage, injuries, or death (Figure 12.2). Effective condition assessment programs prevent these issues and promote more efficient use of government resources, thereby saving taxpayer dollars and improving the lives of citizens.

Appendix A
Sample Cleaning and CCTV Inspection Specification

Sometimes, it can be difficult to create an appropriate specification for a condition assessment program. This appendix contains a sample specification intended for a cleaning and CCTV inspection program provided by a consulting engineering firm. Use this language to help structure your own program, and compare against historical language and suggestions from vendors. This specification language was provided by John C. Mowry, PE of KLH Engineers, Inc., and is reprinted with his permission.

SECTION 02959

CLOSED-CIRCUIT TELEVISION INSPECTION

PART 1: GENERAL

1.01 SUMMARY

A. Section includes internal closed-circuit television (CCTV) inspection of sewers and manholes.

1. Inspect sewers and manholes using color CCTV camera and document inspection on video with audio location and date information, video title information, and continuous tape counter. Provide electronic file and hard copy of inspection logs.
2. The camera, a television monitor, and other components of the video system shall be capable of producing color picture quality to the satisfaction of the OWNER'S REPRESENTATIVE.
3. Schedule CCTV inspection with the OWNER's REPRESENTATIVE.
4. **The CONTRACTOR is to work in conjunction with the OWNER's REPRESENTATIVE during the cleaning/televising process to assign each manhole a designated number. Every manhole is to be assigned an individual number. This designated number is to be used by the CONTRACTOR in the report. At no point is the CONTRACTOR**

to provide a designation without the OWNER's REPRESENTATIVE's approval.

1.02 RELATED SECTIONS

A. Section 02080 Bypassing Sewage

B. Section 02960 Sanitary Sewer Cleaning

1.03 SUBMITTALS

A. Inspection Logs: Unless otherwise indicated, submit inspection logs meeting or exceeding current NASSCO Standards including the following (as a minimum):

1. Project title
2. Time of day
3. Manhole-to-manhole pipe section
4. Pipe segment length
5. Pipe material
6. Line size
7. Compass direction of viewing
8. Direction of camera's travel
9. Pipe depth
10. Operator name
11. Tape counter reading at the beginning and end of each manhole-to-manhole pipe segment
12. Points of infiltration
13. Locations of building sewers
14. Unusual conditions, roots, storm sewer connections, broken pipe, presence of scale and corrosion, and other discernable features.

B. Video record: Submit completed video record after cleaning and rehabilitation.

C. Red-line mark-up maps: In the event the sewer alignment in the field is different than the alignment shown on the map provided with these specifications, the CONTRACTOR shall submit "red-line" - a mark-up map indicating the accurate alignment, manhole location, and manhole designation for all line segments cleaned and CCTV inspected.

D. CONTRACTOR to maintain copy of all inspection documentation (recordings, databases, and logs) for duration of work and warranty period.

PART 2: PRODUCTS

1.04 MATERIALS AND EQUIPMENT

A. Video record: Shall be color format and provided on three (3) appropriately sized external hard drives.

1. Audio portion of composite video record shall be sufficiently free from electrical interference and background noise to provide complete intelligibility of oral report.
2. Store in upright position with temperature range of 45–80°F (7–27°C).
3. Identify each hard drive with labels showing Contract Number, OWNER's name, CONTRACTOR's name, and date(s) work was performed.
4. Identify video files by each manhole-to-manhole pipe segment of sewer line represented, with designations established by OWNER's REPRESENTATIVE.

B. CCTV Inspection Camera(s): Equipped with rotating head, capable of 90° rotation from horizontal and 360° rotation about its centerline.
1. Minimum Camera Resolution: 400 vertical lines and 460 horizontal lines.
2. Camera Lens: Not less than 140° viewing angle, with automatic or remote focus and iris controls.
3. Focal Distance: Adjustable through range of 6 inches (152 mm) to infinity.
4. Camera(s) shall be intrinsically safe and operative in 100% humidity conditions.
5. Lighting Intensity: Remote-controlled and adjusted to minimize reflective glare.
6. Lighting and Camera Quality: Provide clear, in-focus picture of entire inside periphery of sewer.

C. Footage Counter: Measures distance traveled by camera in sewer, accurate to plus or minus 2 feet (0.6 m) in 1000 feet (305 m).

PART 3: EXECUTION

1.05 <u>SEWER FLOW REQUIREMENTS</u>

A. Do not exceed depth of flow shown in Table A.1 for respective pipe sizes as measured in manhole when performing CCTV inspection.

B. When depth of flow at upstream manhole of sewer line section being worked on is above maximum allowable for CCTV inspection, reduce flow to level shown in Table A.1 by pumping and bypassing of flow as specified in Section 02080.

TABLE A.1
MAXIMUM DEPTH OF FLOW FOR TV INSPECTION

Nominal Pipe Diameter	Maximum Depth of Flow
6″-10″	20% of pipe diameter
12″-24″	25% of pipe diameter

1.06 SEQUENCE OF WORK

A. Perform work in the following sequence:
   1. Clean sewer lines and manholes prior to CCTV inspection. Heavy cleaning will be completed as directed by the OWNER's REPRESENTATIVE.
   2. Perform CCTV inspection to comply with requirements of this specification.

1.07 INSPECTION REQUIREMENTS

A. Access: OWNER's REPRESENTATIVE shall have access to observe monitor and other operations at all times.

B. Video Commentary: Record the following information on audio track of video inspection record: narrative of location, direction of view, manhole numbers, pipe diameter and material, date, time of inspection, and location of laterals and other key features.
   1. Video record shall visually display this information at beginning and end of each manhole-to-manhole pipe segment.
   2. Video record between manholes visually displays length in feet from starting point of given segment.
      a. The importance of accurate distance measurements and determining the exact location of service connections is emphasized. Measurement for location of defects shall be above ground by means of a meter device. Marking on the cable, or the like, which would require interpolation for depth of manhole, will not be allowed. Accuracy of the distance meter shall be checked by the use of a walking meter, roll-a-tape, or other suitable devices, and the accuracy shall be satisfactory to the OWNER's REPRESENTATIVE.
   3. Any section of gravity sewer which is found by internal CCTV inspection to be defective, contain silt and/or debris, or be otherwise

unacceptable to the OWNER shall be corrected and re-televised, at the cost of the CONTRACTOR.

C. Sewer Identification: Video record and inspection documentation shall include sewer line and manhole identifiers as established by the OWNER's REPRESENTATIVE. Use upstream manhole as identifier in conjunction with distance meter.
D. Image Perspective: Camera image shall be down center axis of pipe when camera is in motion.
   1. Provide 360° sweep of pipe interior at points of interest, to more fully document existing condition of sewer.
   2. Points of interest may include, but are not limited to the following: defects, encrustations, mineral deposits, debris, sediment, and any location determined not to be clean or part of proper liner installation, and defects in liner that include, but are not limited to, bumps, folds, tears, and dimples.
   3. Cabling system employed to transport camera and transmit its signal shall not obstruct camera's view.
E. Sewer Reach Length: Physically measure and record length of each sewer reach from centerline of its terminal manholes.
F. Inspection Rate: Camera shall be pulled through sewer in either direction, but both inspections are to be in the same direction. **Maximum rate of travel shall be 30 feet (9 m) per minute when recording.**

1.08 FIELD QUALITY CONTROL

A. ENGINEER/OWNER's REPRESENTATIVE will review video record and logs to ensure compliance with requirements listed in this specification.
B. If sewer line, in sole opinion of ENGINEER/OWNER's REPRESENTATIVE, is not adequately clean, it shall be recleaned and CCTV inspected by CONTRACTOR. In the event a segment of interceptor sewer is not sufficiently cleaned after three (3) passes of cleaning **as confirmed by the OWNER's Representative**, the cleaning shall be considered "heavy cleaning" and the CONTRACTOR shall be paid accordingly.
C. All submitted CCTV data shall be reviewed and approved by the ENGINEER prior to payment. Should the data be incomplete in regards to

compliance with the NASSCO Pipeline Assessment and Certification Program (PACP) criteria rating system, the submission will be returned to the CONTRACTOR to complete at no cost to the OWNER or ENGINEER. The CCTV data shall be submitted to the ENGINEER in POSM format for review.

End of Section

SECTION 02960

SANITARY SEWER CLEANING

PART 1: Sewer Line Cleaning

1.01 SCOPE

A. The sewer line cleaning is to remove foreign materials from the lines to allow for unrestricted passage of sewer CCTV equipment.

B. In the event a segment of a sanitary sewer is not sufficiently cleaned after three (3) passes of cleaning **AS CONFIRMED BY THE OWNER's REPRESENTATIVE**, the cleaning shall be considered "heavy cleaning" and the CONTRACTOR shall be paid accordingly. During heavy cleaning if any specialized cutting or cleaning equipment is needed, this shall be included in the heavy cleaning price. While in the heavy cleaning process if the segment being cleaned is not progressing in a reasonable time frame the OWNER's REPRESENATIVE can require the use of specialized cleaning or cutting equipment. The price for this equipment and its use shall be included in the heavy cleaning price.

1.02 PRODUCTS

A. All high-velocity sewer cleaning equipment shall be constructed for ease and safety of operation. The equipment shall have a selection of two or more high-velocity nozzles. The nozzles shall be capable of producing a scouring action from 15 to 45° in all size lines designated to be cleaned. Equipment shall also include a high-velocity gun for washing and scouring manhole walls and floors. The gun shall be capable of producing flows from a fine spray to a solid stream. The equipment shall carry its own water tank, auxiliary engines, pumps, and hydraulically driven hose reel. The equipment shall also include a vacuum, vacuum hose, and enclosed debris tank for appropriate disposal of waste material flushed from the sewer line.

B. Roots shall be removed where root intrusion is a problem. Special attention shall be used during the cleaning operation to assure complete removal of roots from the joints. The use of robotic root-cutting equipment may be used to complete this task.

C. Bucket machines shall be in pairs with sufficient power to perform the work in an efficient manner. Machines shall be belt operated or have an overload device. Machines with direct drive that could cause damage to the pipe will not be allowed. A power rodding machine shall be either a sectional or continuous type capable of holding a minimum of 750 feet of rod. The rod shall be specifically treated steel.

1.03 FIELD PROCEDURES DURING CLEANING OF SEWERS

A. The sanitary sewers to be CCTV inspected shall be cleaned immediately prior to the inspections unless the sewer line walls are sufficiently clean to allow an internal inspection by CCTV to detect structural defects, misalignment, infiltration sources, and root intrusions.

B. All sewer lines and manholes shall be cleaned.

C. During sewer cleaning operations, satisfactory precautions shall be taken in the use of cleaning equipment. When hydraulically propelled cleaning tools, which depend upon water pressure to provide their cleaning force, or tools which retard the flow in the sewer line are used, precautions shall be taken to insure that the water pressure created does not damage or cause flooding of public or private property being served by the sewer.

D. **It shall be the CONTRACTOR's responsibility to provide all water for the hydro-cleaning operation. If the CONTRACTOR wishes to use the local water company as a source of supply for water, he must first contact the water company to determine their permit requirements and/or required fees prior to placing his bid for the work. These costs should be included in the bid price.**

E. The designated sewer sections shall be cleaned using hydraulically propelled, high-velocity jet, or mechanically powered equipment. Selection of the equipment used shall be based on the conditions of lines at the time the work commences. The equipment and methods selected shall be satisfactory to the OWNER and/or the OWNER's REPRESENTATIVE. The equipment shall be capable of removing dirt, grease, rocks, sand, and other materials and obstructions from the sewer lines and manholes.

F. In the event that the CONTRACTOR's equipment becomes lodged in the sewer, the CONTRACTOR shall notify the ENGINEER as soon as practicable, but not later than the end of the same workday. The CONTRACTOR shall take all steps necessary to safely remove the equipment in a timely manner without damaging the sewer, and without causing a sewer overflow. If the lodged equipment must be removed by excavation, the CONTRACTOR shall be responsible for following the OWNER's standards for such excavations and repairs. All such retrieval operations shall be at the CONTRACTOR's sole expense.

G. If roots are encountered during the hydro-cleaning operation which prevents the completion of the cleaning, the CONTRACTOR shall use root-cutting equipment to remove root intrusion as directed by the OWNER's REPRESENTATIVE.

H. All sludge, dirt, sand, rocks, grease, and other solid or semisolid material resulting from the cleaning operation shall be removed at the downstream manhole of the section being cleaned. Passing materials from manhole section to manhole section, which could cause line blockages, accumulations of sand in wet wells, or damage pumping equipment, shall not be permitted.

I. All solids or semisolids resulting from the cleaning operations shall be removed from the site and disposed of at a site approved by the OWNER. All materials shall be removed from the site no less often than at the end of each workday. Under no circumstances will the CONTRACTOR be allowed to accumulate debris, etc., on the work site beyond the stated time, except in totally enclosed containers and as approved by the OWNER. Cost to dispose of material shall be borne by the CONTRACTOR.

J. Acceptance of sewer line cleaning shall be made upon successful completion of the television inspection and shall be to the satisfaction of the OWNER's REPRESENTATIVE. If CCTV inspection shows the cleaning to be unsatisfactory, the CONTRACTOR shall be required to reclean and reinspect the line(s) at the expense of the CONTRACTOR.

End of Section

Index

A

Acoustic detection technologies
- ambient noise reduction, 121
- computing power, 121
- contact microphone, 121
- Echologics, 122
- fire hydrants, 121
- ground microphones, 121
- installation of hydrophone, 119
- leak detection, 122
- listening sticks, 120
- pressure reducing valve, 121
- pressurized water leaks, 119
- size change fixtures, 121

Acoustic fiber optic (AFO) system, 131, 156

Acoustic inspection technologies
- air pockets, 124
- insertion and extraction tube, 125
- *in situ* inspection units, 124
- inspection speed, 123
- linear interpolation, 124
- oil and gas pipelines, 126
- Sahara device, data generated by, 125
- Sahara unit, 123
- SmartBall, 123

Acoustic technologies, 111

Acute exposure guideline levels (AEGL), 98

ALCOSAN, 9, 10

Analog cameras, 36

Analog television and 480i transmission, 36

ArcGIS, 216

Artificial intelligence (AI), 167
- and big data, 226–231
- and machine learning, 138

Asbestos cement (AC), 154

Asset management
- capital budgets, 199
- CCTV-centric tools, 199
- comprehensive asset management program, 198
- condition assessment, 198
- enterprise asset management
  - axioms, 217
  - cloud-based/software-as-a-service platforms, 219
  - computerized maintenance management systems (CMMS), 216
  - condition assessment programs, 216
  - service-level agreement (SLA), 220
- geographic information systems, 209–216
- Government Accounting Standards Board (GASB 34), 62, 200
- manhole inspection programs, 198
- O&M budgets, 199
- planning tools, 205–209
- political support, for water and wastewater, 201

Audio signals, 127

Augmented reality, 233–236

Autonomous vehicles (AVs), 231, 232

Axioms, 217

B

Bacterial and fungal colonization, 19

Battery technology, 139

Bazalgette, Joseph, 4

Big Data, 116–118
- artificial intelligence, 226–231

Biobot Analytics, 186

Broadband electromagnetic inspection (BEM), 163

Bulky equipment, 194

Bump testing, 104

C

Cables, 130

Camera conveyance technology, 27

Cameras, 233

Capacity management operation and maintenance (CMOM) programs, 61

Cast iron, 152, 153

Cast iron pipes, 13, 76

Central processing units (CPUs), 227

Cesspits, 3

Citizens broadband radio service (CBRS), 224

Clay piping, 15

Clean Water Act, 7, 8

Closed circuit television (CCTV) crawlers, 30, 247–248
- contractors, 203
- conventional crawlers, 45
- data integrity, 205
- encourage data entry, 203
- failure of asset, 202
- GIS integration, 204
- initial inspection management, 203
- inspection camera, 249

inspection records, 205
PACP 7.0 standard, 201
slow down delivery time, 203
WinCan, 204
Coding systems, 58, 69
Combined sewers, 7
Comprehensive assessment program, 243
Computerized maintenance management systems (CMMS), 216
Computer software, 58
Concrete
pipes, 18, 151
structure, 17
systems, 17
Condition assessment
buzzwords and jargon, 242
conferences and expositions, 242
data management, 183
detection of stray voltage, 187
errors in assigning files, 183
Global Positioning System (GPS) receivers, 178
inspection hardware, 231–233
PACP coding, 241
RedZone Solo, 182
regular inspection, 188
rim-to-invert measurements, 183
sensor locations, 185
SmartCover, 185
smart infrastructure, 238–240
SSO, 245
stereo cameras, 183
stray voltages, 187
surcharged pipe conditions, 184
survey-grade GPS systems, 179
tamper-detection sensor, 184
telecommunications manholes, 177
3D point cloud, 182
underground power lines, 186
valuable information, 185
visual odometry, 183
Conduits, 1
Consent decrees, 8
Consequence of failure (CoF), 61
Continuous defects, 59
Convolutional neural networks (CNNs), 227
Corrosion model, development of, 109–110
Crawlers, 192
analog cameras, 36
analog television and 480i transmission, 36
CCTV, 30
conventional crawlers, 45
zoom camera and, 35
component and S-video signals, 36
comprehensive condition assessment, 32
conventional camera system, 45
conventional specification, 44
default inspection output format, 43
degrees and pixels, 45
digital technology, 36
digital video, 38
encoders, 46
fiber-optic systems, 46
fisheye cameras, 38, 39, 41
fisheye technology, 43
480p video signal, 37
GIS data or GPS measurements, 46
GoPro cameras, 42
high quality 1080i video, 38
Ibak Panoramo, 182
lighting, 35
lighting quality, 33
municipalities, 34
panning, tilting, and zooming (PTZ), 30, 36
quality of inspection, 32
quantity of pixels, 32
RedZone Robotics, 33, 44
shapes and sizes, 48
side-scan sewer inspection data, 44
size and lighting, 34
skid/floating platform, 34
virtual PTZ inspections, 45
wheel and camera mounting options, 33

D

Data loggers, 106
Deep convolutional neural networks (deep CNNs), 229
Deep manholes, 177
Destructive testing, 159
Diameter interceptors, 23
Dielectric constant, 135, 136
Digital technology, 36
Digital video, 38
Drone-based cameras, 133
Drones, 237
Ductile iron pipe (DIP), 153
Ductile Iron Pipe Research Association (DIPRA), 152
DVD, 78

E

Echologics, 122, 164
Electrochemical sensors, 104
Electrochemical systems, 103
Electromagnetic anomaly detection, 111
Electro Scan, 147
Embedded graphics processors, 237
Encoders, 46
Enhanced mobile broadband (EMBB), 222

Environmental Systems Research Institute (ESRI), 211
European Petroleum Survey Group (EPSG) Registry, 213
Excrement, in Victorian-era London, 5

F

Face ID, 229
Fecal coliform bacteria, 7
Fiber-optic systems, 46
Field quality control, 251–252
Fire hydrants, 121
Fisheye cameras, 38, 39, 41
Fisheye technology, 43
5G and connected platforms
BYOD policy, 226
citizens broadband radio service (CBRS), 224
enhanced mobile broadband (EMBB), 222
high-speed connection, 225
massive machine type communications (MMTC), 224
ultrareliable, low-latency communications (uRLLC), 224
Wi-Fi technology, 222
Fixed point detectors, 103
Flaws, 113
Focused electrode leak location (FELL)
advantage of, 148
dye testing, 144
Electro Scan, 147
external electrode, 146
infiltration, 144
inflows, 144
lining, 145
post-lining acceptance testing, 147
pre-lining FELL inspection, 147
Focused electrodes, 111
Fracta, 167

G

Gas sensor, 103
Gas tracers, 111
flammable gas compounds, 139
helium gas, 141
hydrocarbon sniffers, 140
hydrogen gas molecules, 142
hydrogen gas sensors, 143
leak detection, 141
odorization compounds, 139
smoke testing, 140
sulfur hexafluoride, 141
Germ theory, 5
Global information system (GIS), 52
Global Positioning System (GPS)
inaccurate consumer GPS readings, 179
location and contextual photography, 180
receivers, 178
"survey-grade," 179
survey-grade GPS unit, 179
GoPro cameras, 42
Governmental Accounting Standards Board Statement 34 (GASB 34), 62, 200
Graphics processing units (GPUs), 228
Graphitization, 152
Gravity-fed systems, 1
Gravity sewers, 127
Greenfield Bridge, in Pittsburgh, 12

H

Hacks, 209
High-density polyethylene (HDPE), 16, 156
High-frequency acoustic signals, 69
High quality 1080i video, 38
Hydrogen sulfide-catalyzed corrosion, 19
Hydrogen sulfide gas, 23
Hydrogen sulfide gas corrosion
acute exposure guideline levels (AEGL), 98
bump testing, 104
care and procedures, 97
electrochemical sensors, 104
electrochemical systems, 103
fixed point detectors, 103
gas sensor, 103
health effects, 100
high velocity flows, 97
immediate danger to life and health (IDLH), 102
light sources and photodetector, 103
microbially induced corrosion, 101
neurophysiological abnormalities, 102
optical sensors, 103
oxygen addition and consumption, 99
pH and accelerating corrosion, 99
pipelines and premature failures, 97
regular cleaning, 100
sewer systems, 97
space filling model, 98
T-values, 104
Hydrophone arrays, 130

I

Ibak Panoramo, 182
Illegal connections, 215
Immediate danger to life and health (IDLH), 102
Industrialization, 3
Infrared and multispectral imaging, 111
Infrared lasers, 69

Inspection(s)
hardware, 231–233
preparation and cleaning
contractors, 47, 48
floating systems rugged devices, 49
grit and debris, 47
inspect-to-clean approach, 49
precleaning washes, 47
requirements, 250–251
visual inspections (*see* Visual inspections)
of wastewater infrastructure (*see* Wastewater infrastructure)
waterline, 167–168
Installing meters, 106
International Infrastructure Management Manual, 197
iPad controllers, 30
iPhone, 229
Iron, 151–152
Iron manhole, 173

J

Joint modifier, 55

L

Lampholes, 26, 171
Laser
color-coding scheme, 86
error, 83
floating systems, 83
invisible laser, 82
multiple painstaking measurements, 84
multi-sensor inspection, 84
off-center crawlers, 83
quantitative analysis, 86
RedZone Robotics, 85
safety, 81
sources of error, 81
structured light systems, 82, 84, 85
2D/profiling laser, 85
units, 80
visible light laser, 80
Laterals
crawlers, 192
crossbore inspections, 192
dye tests, 191
LACP inspections, 193
launch cameras, 191
LiquiSmoke, 190
private sewer laterals (PSLs), 188
property owners and municipalities, 189
public/private boundary, 189
smoke tests, 190
standard televising equipment, 191
Lead, 1
Leak detection, 122, 135, 141
Lidar, 231, 232
data, 90
lidar-based systems, 92
MeshLab, 92
newer lidar units, 89
off-the-shelf lidar unit, 88
sources of error, 90
3D reconstruction, 89
units, 87, 88, 91
Lighting, 33, 35
Linear interpolation, 124
LiquiSmoke, 190
Logging device, 106

M

Machine learning (ML), 167
Macomb County, 244
Magnetic flux leakage, 164
Manhole Assessment and Certification Program (MACP) standard
level 1 inspection, 181
level 2 inspection, 181
robots and tools, 182
Manholes
barrel sections, 175
brick and precast manholes, 172
chimney, 175
common construction materials, 172
common pipe sizes and geometries, 172
debris, 170
deep manholes, 177
de facto standard, 171
direction of flow changes, 176
frames, 174, 175
historical application, 169
iron manhole, 173
normal manhole, 177
retractable ladder, 176
round manhole, 173
shallow manholes, 177
square and rectangular covers, 173
stormwater gratings, 174
telecommunications manholes, 173
visual inspection techniques, 175
Manned inspections, 23
Manual of Sewer Condition Classification (MSCC)
chainage and clock position, 52
clock positions, 53
construction material, 52
continuous defect, 55
continuous defects, 54
defects and observations, 53

equipment and practices, 52
global information system (GIS), 52
inspection direction, 52
joint modifier, 55
legal taps, 53
modifier, 55
nutrient-rich sewage, 53
rehabilitation decision, 56
remarks field, 55
sewer inspection crews, 51
water levels, 54
Massive machine type communications (MMTC), 224
MeshLab, 92
Meters and instrument readings
"Big Data" schemes, 116–118
integrated gateways, 115
intelligent compression systems, 116
IoT, 115
long-term storage plan, 116
low-power cellular modems, 115
Small Data, 116
solid-state components, 115
supervisory control and data acquisition (SCADA), 114, 116
Trimble's Telog, 115
Microbially induced corrosion, 17, 101
Microelectronic systems, 139
Microwaves, 135
Microwave sensors, 133
Modifier, 55
Multispectral cameras, 233

N

National Association for Sewer Service Companies (NASSCO), 51, 62, 63, 67, 69, 180, 181, 192
National Pollution Discharge Elimination System (NPDES), 200
Nixon, Lewis, 70
Nondestructive testing methods
broadband electromagnetic inspection, 162
conductive sheet, 162
conventional eddy current testing (ECT), 161
corrosion, 160
electromagnetic field, 162
electromagnetic signals, 163
exciter, 160, 161
failures in pipelines, 160
flux, 160
laser/lidar inspection, 163
magnetic flux leakage, 164
microwave radiation, 165
microwave systems, 165
plastic pipes, 165, 166
tuberculation, 163
Young's modulus, 164
Nonrevenue water, 112
North American Datum of 1983 (NAD 83), 213

O

OdaLogger, 107
Operation and maintenance (O&M) defects, 57
Optical sensors, 103
Original equipment manufacturer, 109
Overall pipe rating (OR), 60

P

Panning, tilting, and zooming (PTZ), 30
Philadelphia, 6
Pipeline Assessment and Certification Program
capacity management operation and maintenance (CMOM) programs, 61
coding systems, 58
computer software, 58
consequence of failure (CoF), 61
continuous defects, 59
Governmental Accounting Standards Board Statement 34 (GASB 34), 62, 200
inspectors and data analysts, 58
lines, 58
operation and maintenance (O&M) defects, 57
overall pipe rating (OR), 60
PACP 7.0, 62–63
pipe rating index (RI), 61
practical issues, of inspection
coding sewer pipelines, 66
coding system, 67
fisheye camera inspections, 66
hydrogen sulfide gas, 67
inspection efficiency, 66
quality control issues, 66
ventilation, 65
video quality, 64
segment grades (SGs), 60
software packages, 60
structural defects, 57
structural grades, 60
Pipelines, 214
Pipe material, 129
Pipe wall measurements
laser and lidar systems, 95
ovality and deflection, 93
source of error, 94
Polyvinyl chloride (PVC), 16, 156
Precipitation events, 6
Premature corrosion, 19
Pressurized water system, 112, 113
Prestressed concrete cylinder pipe (PCCP), 155

Private sewer laterals (PSLs), 188
Profiling sonar, 78
Public safety hazards, 11
Pumping stations, 5
Python Data Analysis Library, 166

R

Radar category, 133
Radar reflections/backscatters, 135
Radar systems, 111
Radio frequency technology, 139
Raster data, 212
Rating index (RI), 61
RedZone Robotics, 33, 44, 85
RedZone's Solo robot, 236
Reliability analysis software, 166
Remotely operated vehicles (ROV), 195
Robotic crawlers, 194
Robots, 237
Roman, engineering techniques, 1
Round manhole, 173
Rust pitting, 13

S

Sahara device 125
Sanitary sewer cleaning, 252–255
Satellite detection, 150
Satellite systems
  dielectric constant, 135, 136
  image processing and technical analysis, 137
  leak detection, 135
  microwaves, 135
  microwave sensors, 133
  points of interest, 138
  radar, 133
  radar reflections/backscatters, 135
  rate of precipitation, 134
  remote sensing of earth, 134
  synthetic aperture radar (SAR) systems, 136
  virtual antenna/aperture, 137
Segment grades (SGs), 60
Self-contained underwater breathing apparatuses (SCUBA), 194
Sewer diving, 26
Sewer flow requirements, 249–250
Sewer systems, 97
Shallow manholes, 177
Sinkholes, 13, 20
Siphons, 1
SL-RAT, 128, 129
Small Data, 116
Smart Cities, 239
SmartCover, 185
Smart infrastructure, 238–240
Smoke tests, 140, 190
Software packages, 62–63
Sonar
  active sonar systems, 71
  cast iron pipes, 76
  debris profile, 75
  debris quantification and ovality measurements, 74
  fan-shaped signal, 78
  flexible pipes, 76
  formats, 78
  fundamental principle of, 72
  higher frequency sounds, 72
  higher-resolution sonar data, 76
  hydrophones, 70, 73
  passive sonar systems, 71
  post processing, 74
  practical considerations, 77
  presentation of data, 75
  profiling sonar, 78
  sell profiling sonar units, 73
  sonar transducer, 72
  TISCIT videos, 78
  ultrasonic waves, 72
  underground pipeline, 71
Sondes, 29
Space filling model, 98
Structural defects, 57
Structural grades, 60
Supervisory control and data acquisition (SCADA), 116
Synthetic aperture radar (SAR) satellites, 136

T

Tamper-detection sensor, 184
Telecommunications manholes, 177
Thermal cameras, 233
Time-of-flight (lidar) systems, 69
Totally Integrated Sonar and Camera Inspection Technique (TISCIT), 78
Transmissive acoustic techniques, 127
T-values, 104

U

Ultrasonic waves, 72
United States Environmental Protection Agency (EPA), 8
Unmanned aerial vehicles (UAVs), 237
Urbanization, 3

V

Vector data, 212, 213
Vendors, 167

Victoria embankment, 4
Video record, 248–249
Virtual reality (VR), 233–236
Visible light ring laser, 69
Visual inspection, 158–159, 175
  camera conveyance technology, 27
  diameter interceptors, 23
  expensive bypass pumping, 23
  high-quality units, 30
  hydrogen sulfide gas, 23
  impressive lighting capabilities, 29
  irregular topography, 25
  “lampholes,” 26
  manned inspections, 23
  optical zooming, 29
  pole and zoom cameras, 29
  push cameras, 27, 28
  recording problems, 23
  rehabilitation and repair systems, 25
  sewer diving, 26
  sondes, 29
Vitrified clay pipes, 15–16

W

Wastewater inspection equipment, 108
Waterline inspection
  long-term consequences, 168
  privatization of water systems, 168
  wastewater system privatization, 168
Water lines
  acoustic fiber optic (AFO) system, 156
  asbestos cement (AC), 154
  cast iron, 152, 153
  chlorine dioxide effects, 158
  concrete pipes, 151
  destructive testing, 159
  ductile iron, 152–153
  ductile iron pipe (DIP), 153
  graphitization, 152
  high-density polyethylene (HDPE), 156
  installation error, 157
  installation issues, 158
  iron, 151–152
  nondestructive testing methods, 160–166
  polymer-based pipelines, 156
  polymers and water treatment systems, 156
  poly-vinyl chloride (PVC), 156
  powder extrusion methodology, 157
  prestressed concrete cylinder pipe (PCCP), 155
  soil and water corrosion, 153
  statistical analyses, 166–167
  visual inspection, 158–159
WGS 84, 213
Wi-Fi technology, 222
Windows Media Player, 78
Wooden sewers, 14

Y

You Only Look Once version 3 (YOLOv3), 230

Z

Zoom camera inspections, 128